THE
MARRIAGE OF
SENSE AND SOUL

THE
MARRIAGE OF
SENSE AND SOUL

INTEGRATING SCIENCE
AND RELIGION

KEN WILBER

RANDOM HOUSE   NEW YORK

LIBRARY OF CONGRESS CATALOGING-IN-PUBLICATION DATA
Wilber, Ken.
The marriage of sense and soul: integrating science and religion /
Ken Wilber.
p.   cm.
Includes bibliographical references and index.
ISBN 0-375-50054-5
1. Religion and science. I. Title.
BL240.2.W526   1998   215—dc21   97-22787

Random House website address: http://www.randomhouse.com/

Manufactured in the United States of America on acid-free paper

24689753

First Edition

CONTENTS

viii / CONTENTS

A NOTE TO THE READER

There is nothing that will cure the senses but the soul,
and nothing that will cure the soul but the senses.

—OSCAR WILDE

It is hard to say exactly when modern science began. Many scholars would date it at roughly 1600, when both Kepler and Galileo started using precision measurement to map the universe. But one thing is certain: starting from whatever date we choose, modern science was, in many important ways and right from the start, deeply antagonistic to established religion.

Most of the early scientists, of course, remained true believers, genuinely embracing the God of the Church; many of them sincerely believed that they were simply discovering God's archetypal laws as revealed in the book of nature. And yet, with the introduction of the scientific method, a universal acid was released that would slowly, inevitably, painfully eat into and corrode the centuries-old steel of religion, dissolving, often beyond recognition, virtually all of its central tenets and dogmas. Within the span of a mere few centuries, intelligent men and women in all walks of life could deeply and profoundly do something that would have utterly astonished previous epochs: deny the very existence of Spirit.

Despite the entreaties of the tenderhearted in both camps, the relation of science and religion in the modern world—that is, in the last three or four centuries—has changed very little since their introduction to each other in the trial of Galileo, where the scientist agreed to shut his mouth and the Church agreed not to burn him. Many wonderful exceptions aside, the plain historical fact has been that orthodox science and orthodox religion deeply distrust, and often despise, each other.

It has been a tense confrontation, a philosophical cold war of global reach. On the one hand, modern empirical science has made stunning and colossal discoveries: the cure of diseases such as typhoid, smallpox, and malaria, which racked the ancient world with untold anguish; the engineering of marvels from the airplane to the Eiffel Tower to the space shuttle; discoveries in the biological sciences that verge on the secrets of life itself; advances in computer sciences that are literally revolutionizing human existence; not to mention plopping a person on the moon. Science can accomplish such feats, its proponents maintain, because it utilizes a solid method for discovering *truth*, a method that is empirical and experimental and based on evidence, not one that relies on myths and dogmas and unverifiable proclamations. Thus science, its proponents believe, has made discoveries that have relieved more pain, saved more lives, and advanced knowledge incomparably more than any religion and its pie-in-the-sky God. Humanity's only real salvation is a reliance on scientific truth and its advance, not a projection of human potentials onto an illusory Great Other before whom we grovel and beg in the most childish and undignified of fashions.

There is a strange and curious thing about scientific truth. As its own proponents constantly explain, science is basically value-free. It tells us what *is*, not what *should be* or *ought to be*. An electron isn't good or bad, it just is; the cell's nucleus is not good or bad, it just is; a solar system isn't good or bad, it just is. Consequently science, in elucidating or describing these basic facts about the universe, has virtually nothing to tell us about good and bad, wise and unwise, desirable and undesirable. Science might offer us truth, but how to use that truth wisely: on this science is, and always has been, utterly silent. And rightly so; that is not its job, that is not what it was designed to do, and we certainly should not blame science for this silence. Truth, not wisdom or value or worth, is the province of science.

In the midst of this silence, religion speaks. Humans seem condemned to meaning, condemned to find value, depth, care, concern, worth, significance to their everyday existence. If science will not (and cannot) provide it, most men and women will look elsewhere. For literally billions of people around the world, religion provides the basic meaning of their lives, the glue of their existence, and offers them a set of guidelines about what is good (e.g., love, care, compassion) and what is not (e.g., lying, cheating, stealing, killing). On the deepest level, religion has even claimed to offer a means of contacting or communing with an ultimate Ground of Being. But by any other name, religion offers what it believes is a genuine *wisdom*.

Fact and meaning, truth and wisdom, science and religion. It is a strange and grotesque coexistence, with value-free science and value-laden religion, deeply distrustful of each other, aggressively attempting to colonize the same small planet. It is a clash of Titans, to be sure, yet neither seems strong enough to prevail decisively nor graceful enough to bow out altogether. The trial of Galileo is repeated countless times, moment to moment, around the world, and it is tearing humanity, more or less, in half.

Fools rush in where angels fear to tread; therefore, the integration of science and religion is the theme of this book. If you are an orthodox religious believer, I would ask only that you relax into the argument and see where it takes you; I do not think you will be dismayed. My primary prerequisite in this discussion is that *both* science and religion must find the argument acceptable in their own terms. For this marriage to be genuine, it must have the free consent of both spouses.

If you are an orthodox scientist, I would only suggest that, as you have a thousand times in the past when you were working on a problem, let curiosity and wonder bubble up, but in this case don't focus it on a specific solution. Simply let wonder fill your being until it takes you out of yourself and into the staggering mystery that is the existence of the world, a mystery that facts alone can never begin to fill. If Spirit does exist, it will lie in that direction, the direction of wonder, a direction that intersects the very heart of science itself. And you will find, in this adventure, that the scientific method will never be left behind in the search for an ultimate ground.

And we all know how to wonder, don't we? From the depths of a Kosmos too miraculous to believe, from the heights of a universe too wondrous to worship, from the insides of an astonishment that has no boundaries, an answer begins to suggest itself, and whispers to us lightly. If we listen very carefully, from within this infinite wonder, perhaps we can hear the gentle promise that, in the very heart of the Kosmos itself, both science and religion will be there together to welcome us home.

K.W.
BOULDER, COLORADO
SUMMER 1997

PART I

THE PROBLEM

1

THE CHALLENGE OF OUR TIMES: INTEGRATING SCIENCE AND RELIGION

There is arguably no more important and pressing topic than the relation of science and religion in the modern world. Science is clearly one of the most profound methods that humans have yet devised for discovering *truth*, while religion remains the single greatest force for generating *meaning*. Truth and meaning, science and religion; but we still cannot figure out how to get the two of them together in a fashion that *both* find acceptable.

The reconciliation of science and religion is not merely a passing academic curiosity. These two enormous forces—truth and meaning—are at war in today's world. Modern science and premodern religion aggressively inhabit the same globe, each vying, in its own way, for world domination. And something, sooner or later, has to give.

Science and technology have created a global and transnational framework of industrial, economic, medical, scientific, and informational systems. Yet however beneficial those systems may be, they are all, in themselves, devoid of meaning and value. As its own proponents constantly point out, science tells us what is, not what should be. Science tells us about electrons, atoms, molecules, galaxies, digital data bits, network systems: it tells us what a thing is, not whether it is good or bad, or what it should be or could be or ought to be. Thus this enormous global scientific infrastructure is, in itself, a value-less skeleton, however functionally efficient it might be.

Into this colossal value vacuum, religion has happily rushed. Science has created this extraordinary worldwide and global framework—itself utterly devoid of meaning—but within that ubiquitous framework, subglobal pockets of premodern religions have created value and meaning for billions of people in every part of the world. And these same premodern religions often deny validity to the scientific framework within which they live, a framework that provides most of their medicine, economics, banking, information networks, transportation, and communications. Within the scientific skeleton of truth, religious meaning attempts to flourish, often by denying the scientific framework itself—rather like sawing off the branch on which you cheerily perch.

The disgust is mutual, because modern science gleefully denies virtually all of the basic tenets of religion in general. According to the typical view of modern science, religion is not much more than a holdover from the childhood of humanity, with about as much reality as, say, Santa Claus. Whether the religious claims are more literal (Moses parted the Red Sea) or more mystical (religion involves direct spiritual experience), modern science denies them all, simply because there is no credible empirical evidence for any of them.

So here is the utterly bizarre structure of today's world: a scientific framework that is global in its reach and omnipresent in its information and communication networks, forms a meaningless skeleton within which hundreds of subglobal, premodern religions create value and meaning for billions; and they each—science and religion each—tend to deny significance, even reality, to the other. This is a massive and violent schism and rupture in the internal organs of today's global culture, and this is exactly why many social analysts believe that if some sort of reconciliation between science and religion is not forthcoming, the future of humanity is, at best, precarious.

WHAT DO WE MEAN BY "RELIGION"?

The aim of this book is to suggest how we might begin to think about both science and religion in ways that allow their reconciliation and eventual integration, *on terms acceptable to both parties.*

Of course, this reconciliation of science and religion depends, in part, on exactly what we mean by "science" and "religion." We will actually devote several chapters to just this topic (Chapters 11, 12, and 13). In the meantime, a few crucial points should be noted.

Defining "religion" is itself an almost impossible task, largely because there are so many different forms of the beast that it becomes hard to spot what, if anything, they have in common. But one thing is immediately obvious: many of the specific and central claims of the world's great religions *contradict each other, but if we cannot find a common core of the world's great religions, then we will never find an integration of science and religion.*

Indeed, if we cannot find a common core that is generally acceptable to most religions, we would be forced to choose one religion and deny importance to the others; or we would have to "pick and choose" tenets from among various religions, thus alienating the great religious traditions themselves. We would never arrive at an integration of science and religion that both parties would find acceptable, because most religions would reject what was done to their beliefs in order to force this "reconciliation."

It will do no good, for example, to claim, as many Christian creationists have, that the Big Bang suggests that the world is the product of a personal creator God, when one of the most profound and influential religions in the world, Buddhism, does not believe in a personal God to begin with. Thus, we cannot use the Big Bang in order to "integrate" science and religion unless we can first find a way to reconcile Christianity and Buddhism (and the world's wisdom traditions in general). Otherwise, we are not integrating science and religion; we are simply "integrating" one narrow version of Christianity with one version of science. This is not worthy of the term "integration," and it is certainly not an integration that other religions would find acceptable.

Thus, those who wish to advocate one particular form of religion—whether it be a patriarchal God the Father, a matriarchal Great Goddess, a fundamentalist Christianity, a mythological Shintoism, a Gaia ecoreligion, a fundamentalist Islam—have often taken various modern developments in science and attempted to show that these developments just happen to fit with a very generous interpretation of their particular religion. This will not be our approach. Because the fact is, unless science can be shown to be compatible with certain deep features common to *all* of the world's major wisdom traditions, the long-sought reconciliation will remain as elusive as ever.

So before we can even attempt to integrate science and religion, we need to see if we can find a common core of the world's great wisdom traditions. This common core would have to be a general frame that, shorn of specific details and concrete contents, would nonetheless be acceptable to most religious traditions, at least in the abstract. Is there such a common core?

The answer, it appears, is yes.

THE GREAT CHAIN OF BEING

Huston Smith—whom many consider the world's leading authority on comparative religion—has pointed out, in his wonderful book *Forgotten Truth*, that virtually all of the world's great wisdom traditions subscribe to a belief in the Great Chain of Being. Smith is not alone in this conclusion. From Ananda Coomaraswamy to René Guénon, from Fritjof Schuon to Nicholas Berdyaev, from Michael Murphy to Roger Walsh, from Seyyed Nasr to Lex Hixon, the conclusion is consistent: the core of the premodern religious worldview is the Great Chain of Being.

According to this nearly universal view, reality is a rich tapestry of interwoven levels, *reaching from matter to body to mind to soul to spirit*. Each senior level "envelops" or "enfolds" its junior dimensions—a series of nests within nests within nests of Being—so that every thing and event in the world is interwoven with every other, and all are ultimately enveloped and enfolded by Spirit, by God, by Goddess, by Tao, by Brahman, by the Absolute itself.

As Arthur Lovejoy abundantly demonstrated in his classic treatise on the Great Chain, this view of reality has in fact "been the dominant official philosophy of the larger part of civilized humankind through most of its history." The Great Chain of Being is the worldview that "the greater number of the subtler speculative minds and of the great religious teachers [both East and West] have, in their various fashions, been engaged in." This stunning unanimity of deep religious belief led Alan Watts to state flatly that "We are hardly aware of the extreme peculiarity of our own position, and find it difficult to realize the plain fact that there has otherwise been a single philosophical consensus of universal extent. It has been held by [men and women] who report the same insights and teach the same essential doctrine whether living today or six thousand years ago, whether from New Mexico in the Far West or from Japan in the Far East."

The Great Chain of Being—that is perhaps a bit of a misnomer, because, as I said, the actual view is more like the Great Nest of Being, with each senior dimension enveloping or enfolding its junior dimension(s)—a situation often described as "transcend and include." Spirit transcends but includes soul, which transcends but includes mind, which transcends but includes the vital body, which transcends but includes matter. This is why the Great Nest is most accurately portrayed as a series of concentric spheres or circles, as I have indicated in Figure 1-1.

This is not to say that every single religious tradition from time immemorial has possessed exactly this particular scheme of matter, body, mind, soul, and spirit; there has been considerable variation within it. Some traditions have only three basic levels in the Great Nest—usually body, mind, and spirit. As Chögyam Trungpa, Rinpoche, pointed out in *Shambhala: The Sacred Path of the Warrior*, this simple hierarchy of body, mind, and spirit was nonetheless the backbone

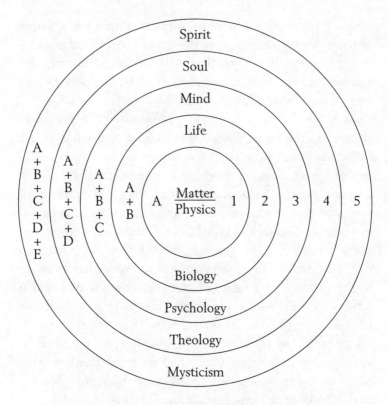

FIGURE 1-1. THE GREAT NEST OF BEING

of even the earliest shamanic traditions, showing up as the hierarchy of earth, human, and heaven. This three-level scheme reappears in the Hindu and Buddhist notion of the three great states of being: *gross* (matter and body), *subtle* (mind and soul), and *causal* (spirit). Many of these traditions, on the other hand, also have extensive subdivisions of the Great Nest, sometimes breaking it down into five, seven, twelve, or even more levels and sublevels.

But the basic point has remained essentially identical: Reality is a series of nests within nests within nests, reaching from matter to mind to Spirit, with the result that all beings and all levels were ultimately enfolded in the all-pervasive and loving embrace of an ever-present Spirit.

Each senior level in the Great Nest, although it includes its juniors, nonetheless possesses emergent qualities not found on the junior level. Thus, the vital animal body *includes* matter in its makeup, but it also *adds* sensations, feelings, and emotions, which are not found in rocks. While the human mind *includes* bodily emotions in its makeup, it also *adds* higher cognitive faculties, such as reason and logic, which are not found in plants or other animals. And while the soul *includes* the mind in its makeup, it also *adds* even higher cognitions and affects, such as archetypal illumination and vision, not found in the rational mind. And so on.

In short, each higher level possesses the essential features of its lower level(s), but then adds elements not found on those levels. Each higher level, that is, *transcends* but *includes* its juniors. And this means that each level of reality has a different architecture, so to speak.

For just that reason, each level of reality, according to the great traditions, has a specific branch of knowledge associated with it (which I have also indicated in Figure 1-1): Physics studies matter. Biology studies vital bodies. Psychology and philosophy address the mind. Theology studies the soul and its relation to God. And mysticism studies the formless Godhead or pure Emptiness, the radical experience of Spirit beyond even God and the soul.

Such has been the dominant worldview, in one variation or another, for most of humankind's history and prehistory. It is the backbone of the "perennial philosophy," the nearly universal consensus about reality held by humanity for most of its time on this earth. Until, that is, the rise of modernity in the West.

THE MODERN DENIAL
OF SPIRITUALITY

With the rise of modernity in the West, the Great Chain of Being almost entirely disappeared. As we will see, the modern West, after the Enlightenment, became the first major civilization in the history of humanity to deny almost entirely the existence of the Great Nest of Being.

In its place was a "flatland" conception of the universe as composed basically of matter (or matter/energy), and this material universe, including material bodies and material brains, could best be studied by science, and science alone. Thus, in the place of the Great Chain reaching from matter to God, there was now matter, period. And so it came to pass that the worldview known as *scientific materialism* became, in whole or part, the dominant official philosophy of the modern West.

Many religiously minded scholars have noted this modern "collapse" of the Great Nest of Spirit, and lamented it loudly. They have blamed this collapse on everything from the Newtonian-Cartesian paradigm to patriarchal domination, from capitalistic commodification of life's values to anti-Goddess male aggression, from hatred of the holistic web of life to a devaluation of nature in favor of analytic abstractions, from material lust and greed to obsession with monetary gain. The list of malevolent causes is indeed virtually endless.

True as those explanations might be, none of them addresses the core issues. As we will see, there is good reason that the Great Chain in its traditional form collapsed. The Great Nest of Spirit simply could not stand up to certain undeniable truths ushered in by modernity, and if we are to integrate both premodern religion and modern science, the truths of both parties must be brought to the union. And modernity possessed a tremendous share of new truths and new discoveries—it was far from being the Great Satan.

At the same time, the rise of modernity was beset with its own grave problems, not the least of which was the massive cultural earthquake brought about by the shuddering collapse of the Great Nest of Spirit. No longer were men and women enfolded in Spirit, they were simply awash in matter: hardly a comforting universe.

So we reach a crucial point. Our aim is to integrate premodern religion with modern science. We have already seen that the core of premodern religion is the Great Nest of Being. But what exactly is the core of modernity? *If we are to integrate premodern and modern, and if premodern is the Great*

Chain, then what exactly is "modern"? The key to the long-sought integration might very well lie in this neglected direction.

WHAT IS "MODERNITY"?

What specifically did modernity bring into the world that the premodern cultures by and large lacked? What made modernity so substantially *different* from the cultures and epochs that preceded it? Whatever it was, it will have to be an essential feature of the sought-for integration.

Many answers have been offered to the question "What is modernity?" Most of them are decidedly negative. Modernity, it is said, marked the death of God, the death of the Goddess, the commodification of life, the leveling of qualitative distinctions, the brutalities of capitalism, the replacement of quality by quantity, the loss of value and meaning, the fragmentation of the lifeworld, existential dread, a rampant and vulgar materialism—all of which have often been summarized in the phrase made famous by Max Weber: "the disenchantment of the world."

No doubt there is some truth to all those claims, and we will give them abundant consideration. But clearly modernity has some immensely positive aspects as well, for it also gave us the liberal democracies; the ideals of equality, freedom, and justice, regardless of race, class, creed, or gender; modern medicine, physics, biology, and chemistry; the end of slavery; the rise of feminism; and the universal rights of humankind. Those, surely, are a little more noble than the mere "disenchantment of the world."

No, we need a specific definition or description of modernity that allows for all those factors, both good (such as liberal democracies) and bad (such as the widespread loss of meaning). Various scholars, from Max Weber to Jürgen Habermas, have suggested that what specifically defines modernity is something called "the differentiation of the cultural value spheres," which especially means the differentiation of art, morals, and science. Where previously these spheres tended to be fused, modernity differentiated them and let each proceed at its own pace, with its own dignity, using its own tools, following its own discoveries, unencumbered by intrusions from the other spheres.

This differentiation allowed each sphere to make profound discoveries that, if used wisely, could lead to such "good" results as democracy, the end of slavery, the rise of feminism, and rapid advances in medical science; but discoveries that, if used unwisely, could just as easily be perverted into the "downsides" of modernity, such as scientific imperialism, the disenchantment of the world, and totalizing schemes of world domination.

The brilliance of this definition of modernity—namely, that it differentiated the value spheres of art, morals, and science—is that it allows us to see the underpinnings of *both* the good news and the bad news of modernity. In ways that will become more obvious in the following chapters, this definition allows us to understand both the *dignity* and the *disaster* of modernity, and we will explore each of them very carefully.

Premodern cultures certainly possessed art, morals, and science. The point, rather, is that these spheres tended to be relatively "undifferentiated." To give only one example now, in the Middle Ages, Galileo could not freely look through his telescope and report the results because art and morals and science were all fused under the Church, and thus the morals of the Church defined what science could—or could not—do. The Bible said (or implied) that the sun went around the earth, and that was the end of the discussion.

But with the differentiation of the value spheres, a Galileo could look through his telescope without fear of being charged with heresy and treason. Science was free to pursue its own truths unencumbered by brutal domination by the other spheres. Likewise with art and morals: Artists could, without fear of punishment, paint nonreligious themes, or even sacrilegious themes, if they wished. And moral theory was free to pursue an inquiry into the good life, whether it agreed with the Bible or not.

For all those reasons and more, these *differentiations of*

modernity have also been referred to as the *dignity* of modernity, for these differentiations were in part responsible for
the rise of liberal democracy, the end of slavery, the growth
of feminism, and the staggering advances in the medical sciences, to name but a few of these many dignities.

As we will see, the "bad news" of modernity was that these
value spheres did not just peacefully separate, they often
flew apart completely. The wonderful *differentiations* of
modernity went too far into actual *dissociation*, fragmentation, alienation. Dignity became disaster. The growth became
a cancer. As the value spheres began to dissociate, this allowed a powerful and aggressive science to begin to invade
and dominate the other spheres, crowding art and morals out
of any serious consideration in approaching "reality." Science
became *scientism*—scientific materialism and scientific imperialism—which soon became the dominant "official" worldview of modernity.

It was this scientific materialism that very soon pronounced the other value spheres to be worthless, "not scientific," illusory, or worse. And for precisely that reason, it was
scientific materialism that *pronounced the Great Chain of
Being to be nonexistent.*

According to scientific materialism, the Great Nest of
matter, body, mind, soul, and spirit could be thoroughly and
rudely reduced to systems of matter alone; and matter—
whether in the material brain or material process systems—
would account for all of reality, without remainder. Gone
was mind and gone was soul and gone was Spirit—gone, in
fact, was the entire Great Chain, except for its pitiful bottom
rung—and in its place, as Whitehead famously lamented,
there was reality as "a dull affair, soundless, scentless, color-
less; merely the hurrying of material, endlessly, meaning-
lessly."

And so it came about that the modern West was the first
major civilization in the history of the human race to deny
substantial reality to the Great Nest of Being. It is into this
massive and universal denial that we wish to attempt to reintroduce the spiritual dimension, but on terms acceptable to
science as well.

CONCLUSION

To integrate religion and science is to integrate a premodern worldview with a modern worldview. But we saw that the essence of premodernity is the Great Chain of Being, and the essence of modernity is the differentiation of the value spheres of art, morals, and science. Thus, in order to integrate religion and science, we need to *integrate the Great Chain with the differentiations of modernity.* As we will start to see in the next chapter, this means that each of the levels in the traditional Great Chain needs to be carefully differentiated in the light of modernity. If we can do that, we will have satisfied *both* the core claim of spirituality—namely, the Great Chain—and the core claim of modernity—namely, the differentiation of the value spheres.

If this integration can be done without "cheating"—that is, without stretching and deforming either religion or science to a point where they do not recognize themselves—then this will be an integration that both parties can genuinely embrace. Such a synthesis would unite the best of premodern wisdom with the brightest of modern knowledge, bringing together truth and meaning in a way that has thus far eluded the modern mind.

———⚭———

2

A DEADLY DANCE:
THE RELATION OF SCIENCE AND
RELIGION IN TODAY'S WORLD

With the rise of modernity—and the collapse of the Great Nest of Being—science and religion began an antagonistic dance. Perhaps it would be more accurate to say that science and religion entered into a fierce and complex war, a war between epochs, a war between worlds, a war between a premodern and mythological orientation to the universe and a thoroughly tough-minded and modern gaze, rational in its aspirations.

In the wake of modernity—in the wake, that is, of the multifarious events generally associated with the eighteenth-century Enlightenment and often continuing to today (events we will examine in Chapter 4)—there arose four or five major stances toward the relation of science and religion. These stances are still with us, and they still dominate the discussion about science and spirituality. Yet all of them have substantial, even severe, limitations. Nonetheless, we can learn much from both their strengths and their weaknesses, their contributions and their flaws. In particular, understanding why these attempts have largely failed, and continue to fail, will help us zero in on the precise requirements of this difficult marriage.

1. Science denies any validity to religion. ⟶ This is the standard empirical and positivist approach, which became, in numerous guises, the dominant official mood of modernity. Classic variations on this theme have been given

by Auguste Comte, Sigmund Freud, Karl Marx, and Bertrand Russell, but they all boil down to: religion is a hangover from the childhood of humanity, on exactly the same footing as the tooth fairy. It's cute for kids but deadly for adults, and its persistence into maturity—the persistence of deeply held religious beliefs into adulthood—is a sign of pathology, lack of logical clarity, or existential inauthenticity. There are no exceptions, because there is no God. And there is no God because science registers that which is real, and no microscope and no telescope have yet spotted any "God."

2. Religion denies any validity to science. ∞ This is a typically fundamentalist retort to modernity, and is itself a by-product of modernity. By and large, classical religions *never* denied science—first, because science was not a threat (only with modernity does science become powerful enough to kill God); and second, because science was always held to be one of several valid modes of knowing, subservient to spiritual modes but valid nonetheless, and hence there was no reason to deny its importance.

At the same time, it should be noted that the sciences of antiquity were not nearly as impressive as what a Newton, a Galileo, or a Kepler would deliver, and thus few were the temptations to make science itself into a new religion of positivism (which is exactly what Auguste Comte would propose, with Comte himself volunteering to be, literally, the Pope of Positivism).

In any event, science in the premodern world had little inclination to deny religion and thus a drastic counterforce was uncalled for. But with the rise of modernity and its inherent claim that all religions are childish productions, many fundamentalist religions (especially Christianity and Islam) began to deny even the basic facts of science itself: evolution does not exist, the Earth was literally created in six days, radiocarbon dating is a fraud, and so on. It has been pointed out, for example, that the extremism of Islamic fundamentalists is not so much an inherent aspect of Islam (which has produced some truly glorious civilizations) as it is a product of a wild counterreaction to modernity's attempt to terrorize and kill spirituality in general. In wild panic, the fundamentalists have become counterterrorists.

This does not in any way excuse terrorism; I believe that many (but by no means all) of the religious sentiments of humankind are indeed a childish hangover and eventually need to be surrendered. Most fundamentalists, in this sense, are indeed refusing to grow up cognitively. But it does point out the intense emotions involved in this battle of modernity, this battle to find a place for both science and religion, truth and meaning, logic and God, facts and Spirit, evidence and the eternal.

3. Science is but one of several valid modes of knowing, and thus can peacefully coexist with spiritual modes. ∞ This was the standard position of most classical religions and the religions of antiquity. In fact, this is just another way of describing the Great Chain of Being, and, as you can see in Figure 1-1, science and theology and mysticism all had an important and rightful place in the Great Nest of Being.

Although this view—which is now generally called *epistemological pluralism*—was the backbone of the great wisdom traditions, it collapsed with the Great Chain, upon which it depended. *When modernity rejected the Great Chain, it simultaneously rejected epistemological pluralism.* And modernity continues to reject epistemological pluralism in any of its forms because modernity most definitely continues to reject the Great Chain.

Nonetheless, for those scholars, theorists, and intelligent laypeople attempting to make sense of the universe in some sort of holistic or encompassing fashion, epistemological pluralism has remained one of the more appealing and sophisticated attempts to unite science and religion. Modernity itself aggressively discarded this view; but what we might call "countermodernity" or the "counterculture"—never much more than a small percentage of the total population, but one desperately looking for a way to heal modernity's frag-

mentations—would nevertheless continue to look to epistemological pluralism as one of the most refined and compelling ways to proceed.

The traditional view of epistemological pluralism was
given perhaps its clearest statement by such Christian mystics as St. Bonaventure and Hugh of St. Victor: every human
being has the eye of flesh, the eye of mind, and the eye of
contemplation. Each of these modes of knowing discloses its
own corresponding dimension of being (gross, subtle, and
causal), and thus each is valid and important when addressing its own realm. This gives us a balance of empirical knowledge (science), rational knowledge (logic and mathematics),
and spiritual knowledge (gnosis).

The three eyes of knowing are, of course, just a simplified
version of the universal Great Chain of Being. If we picture
the Great Chain as having five levels (matter, body, mind,
soul, and spirit), men and women have five eyes available to
them (material prehension, bodily emotion, mental ideas, the
soul's archetypal cognition, and spiritual gnosis). Likewise, if
the Great Chain is divided into twelve levels, we have twelve
eyes, or twelve levels of awareness and knowing.

Indeed, Plotinus—arguably the greatest philosopher-
mystic the world has ever known—usually gave the Great
Chain twelve levels: matter, life, sensation, perception, impulse, images, concepts, logical faculty, creative reason, world
soul, nous, and the One. Table 2-1 shows the typical Great
Nest as presented by Plotinus and Sri Aurobindo, two of its
greatest representatives (the match between them is quite
striking and fairly typical).

The point is that, any way we slice the great pie—three
levels, five levels, twelve levels or more—men and women
have available to them *at least* the three basic eyes of knowing: the eye of flesh (empiricism), the eye of mind (rationalism), and the eye of contemplation (mysticism), each of
which is important and quite valid when dealing with its
own level, but gravely confused if it attempts to see into
other domains. This is the very heart of epistemological pluralism, and, as far as it goes, it is indeed quite valid.

Now, if the existence of all three eyes of knowing were a

TABLE 2-1. THE GREAT NEST ACCORDING TO PLOTINUS AND AUROBINDO

PLOTINUS	AUROBINDO
Absolute One (Godhead)	Satchitananda/Supermind (Godhead)
Nous (intuitive mind) (subtle)	Intuitive mind/Overmind
Soul/World Soul (psychic)	Illumined World mind
Creative Reason (vision-logic)	Higher mind/Network mind
Logical faculty (formop)	Logical mind
Concepts and opinions	Concrete mind (conop)
Images	Lower mind (preop)
Pleasure/pain (emotions)	Vital-emotional; impulse
Perception	Perception
Sensation	Sensation
Vegetative life function	Vegetative
Matter	Matter (physical)

commonly accepted fact in modernity, the relation of science and religion—and their peaceful coexistence—would be no problem whatsoever. Empirical science would pronounce on the facts delivered by the eye of flesh, and religion would pronounce on the facts delivered by the eye of spirit (or the eye of contemplation). But mainstream modernity has soundly and thoroughly denied reality to the eye of spirit. Modernity recognizes only the eye of reason yoked to the eye of flesh—in Whitehead's phrase, the dominant worldview of modernity is *scientific materialism*, and whether that science be the holistic science of systems theory or the subatomic physics of quantum events, science is the eye of reason linked to evidence offered by the empirical senses. *In no case is the eye of contemplation or the eye of Spirit required . . .* or even allowed.

The real difficulty, then, is not showing how empiricism, rationalism, and mysticism can all fit together in the Great Chain of Being; it is not showing how they can all be harmoniously integrated in a great spectrum of consciousness; it is not demonstrating that such a synthesis is coherent and complete. For *that*, in a sense, is the easy part. All of those statements, I believe, are true. The hard part is that modernity *does not accept the higher levels themselves* (the transmental, transrational, transpersonal, and contemplative modes), and thus it sees *no need* whatsoever for the integration. Why try to integrate science and Santa Claus?

Thus, arguing for epistemological pluralism and the different eyes of knowing (or modes of inquiry) is at best a first step. The real problem is that modernity does not accept the terms to be integrated in the first place. We will therefore have to find another path into the heart of modernity if an integration of science and religion is ever to take root in the West.

4. Science can offer "plausibility arguments" for the existence of Spirit. ∞ This is a variation on epistemological pluralism, but because it has recently generated much interest—among professionals and laypeople alike—I will discuss it separately. The idea is that, as empirical science pushes into the deepest secrets of the physical world, it discovers facts and data that seem to demand some sort of Intelligence beyond the material domain.

The standard example is the Big Bang: Where did *that* come from? Since the very earliest material plasma seems to have been obeying mathematical laws that themselves did not come into being with the Big Bang, must not those laws exist "in the mind of some eternal Spirit," as Sir Arthur Eddington, echoing Berkeley, suggested? These laws, all agree, existed prior to space and time. Thus, to the question "What existed before the Big Bang?," the answer very well might be *a nonmaterial Logos governing the patterns of creation*—what many would simply call God. And, this argument continues, since science discovered the Big Bang, science itself is pointing to God.

There are numerous variations on this argument, most of which are twists on the traditional *argument from design*, namely, that incredibly intelligent natural designs demand an incredibly intelligent Something-or-Other behind them. This is a very old argument, stretching back at least to early Greece, that has been aggressively attached to recent advances in the sciences—particularly quantum, relativistic, systems and complexity theories.

This approach is perhaps the simplest and most popular of the ways in which an alienated countermodernity has attempted to integrate science and religion. We see it in everything from *The Tao of Physics* (which maintains that modern physics discloses a worldview similar to that of Eastern mysticism) to the thoughtful writings of Paul Davies (e.g., *The Mind of God*, which maintains that "By the means of science we can truly see into the mind of God") to the Anthropic Principle (which maintains that the evolution of human beings is so improbable that the universe must have known what it was doing from the very start) to the "new holistic paradigm" approaches (which maintain that systems theory is demonstrating the same great web of life that the holistic spiritual traditions embraced).

I have a great deal of sympathy for many of those plausibility arguments. They are suggestive. They are indicative. They are certainly entertaining. But, alas, none of them can stand up to the critical philosophy of, say, Immanuel Kant or the Buddhist genius Nagarjuna, both of whom strongly demonstrated the limits of rationality in the face of the Divine. If deeply spiritual Nagarjuna is unswayed by these plausibility arguments, how must they go over with nonspiritual types? This is why the vast majority of scientists—and modernity itself—tend to take these "arguments" with mild interest at best, wild amusement at worst.

The real problem with these rational, mental, or linguistic plausibility arguments is that they are *attempting to use the eye of mind to see that which can be seen only with the eye of contemplation.* This confusion of levels (which is called a "category error") is particularly fatal when it comes to "arguments" for the existence of Spirit. It was precisely these inadequate and altogether unconvincing mental attempts to storm the spiritual palace that made modernity look with suspicion on any and all claims to be able to prove the existence of God. These arguments and "proofs" are simply not compelling to the modern mind, and to the spiritual mind they are inherently inadequate anyway. In no case, then, do these arguments deliver what they aspire to deliver, namely, any sort of actual spiritual knowledge.

Martin Gardner's response to these arguments is quite typical. Referring to the Anthropic Principle, Gardner points out that, according to its adherents, it comes in four successive forms, each stronger in its claims: the Weak Anthropic Principle, or WAP (the universe allows us to exist); the Strong Anthropic Principle, or SAP (the existence of life explains the laws of the universe); the Participatory Anthropic Principle, or PAP (conscious observers are necessary to bring the universe into existence); and the Final Anthropic Principle, or FAP (if life or consciousness ends, the universe will evaporate).

To this list Gardner, speaking for modernity, adds the Completely Ridiculous Anthropic Principle, or CRAP: any who buy the first four.

5. Science itself is not knowledge of the world but merely an interpretation of the world, and therefore it has the same validity—no more, no less—as poetry and the arts. ∞ Because science refused to gracefully take its place as one among many other valid modes of knowing, this approach attempted to cripple science in its very foundations, pulling it down against its will. It tried to level the playing field by shooting science in the head and proclaiming, "There! Now we're equal."

This approach is, of course, the essence of postmodernism. It says, in effect, that the world is not *perceived*, it is only *interpreted*. Different interpretations are equally valid ways of making sense of the world, and thus no interpretation is intrinsically better than another. Science is not a privileged conception of the world but merely one among many equivalent interpretations; science does not offer "truth" but simply its own favorite prejudice; science is not a set of universal facts but merely an arbitrary imposition of its own power drives. And in all cases, science is no more grounded in reality than is any other interpretation, so that, epistemologically speaking, there is little difference between science and poetry, logic and literature, history and mythology, fact and fiction.

Thus, this postmodern view continues, science is not gov-

erned by *facts*, it is governed by *paradigms*, and paradigms are
not much more than ad hoc constructions or free-floating interpretations. As we will see in the next chapter, the notion
that science is governed by paradigms was made popular by
Thomas Kuhn in his now-famous *The Structure of Scientific
Revolutions*, which the postmodernists seized with a fury. Yet
this is not at all the way Kuhn defined or described paradigms, and he strenuously denounced this abuse of his
work—to no avail. But, according to this mis-Kuhnian notion
of "paradigm," science is not conforming itself to actual facts,
it is simply imposing its paradigms on the world at large.
Since independent facts do not exist (only interpretations
do), it follows, according to this account, that science is always driven by some sort of power or ideology: science itself
is sexist, racist, ethnocentric, imperialistic, brutally imposing
its analytic and divisive interpretations on an unwilling and
innocent world. And it has no more warrant, no more final
validity, than any other poetic interpretation.

Science reduced to poetry: *this is now the dominant route
taken by the countercultural world in an attempt to reduce the
monster of science to manageable proportions.* The postmodern
attack on science, which attempts to shatter its epistemological foundations, is the reigning model of how to "counter"
science in the postmodern world.

It is crucial to understand this postmodern attempt to
level science and thus make room for "other paradigms,"
whether poetry, religion, mysticism, astrology, holism, poststructuralism, neopaganism, or whatnot. Aside from its important moments of truth (which I will highlight and
incorporate), this entire postmodern attempt is nonetheless
profoundly misguided and deeply confused. In attempting to
shoot science in the head, it simply kills that which it should
be integrating. It denies that which should be embraced. It
sabotages the desired wedding by murdering one of the
spouses.

Yet this postmodern attempt—to see science as paradigm-
bound, and thus rush in to offer a "new paradigm"—is at the
core of virtually every alternative, countercultural, and "new-
paradigm" approach to science and religion. The idea is that

science is undergoing a paradigm shift of momentous proportions, and this new paradigm is in fact commensurate with spiritual realities. The "new paradigm," it is claimed, will therefore unite science and religion for the first time in history, thus heralding the beginning of a world transformation, global in its sweep, that will usher in the beginning of a holistic, unified, web-of-life world.

And almost all of that, we will see, is based on a complete misreading of Thomas Kuhn.

A POSSIBLE SOLUTION

In the following chapters, I will argue that all five of those stances toward science and religion—and their possible integration—are inadequate. The first two stances—science denies religion, religion denies science—are obviously not going to be integrative. But the other three (epistemological pluralism, plausibility arguments, and postmodern/paradigm) have not proven powerful enough to integrate science and religion in a fashion that *both* parties find acceptable.

We have seen that the only way we can possibly integrate science and religion is by integrating the Great Chain with the major differentiations of modernity. *This specific integration is where all of these stances tend to fail,* yet it is this approach that very likely contains the central solution.

Let us use, as a brief example, epistemological pluralism—which, as we saw, is by far the most sophisticated of the alternatives. In the traditional view of epistemological pluralism, science is placed *on the bottom rung* of the great hierarchy. In this view, recall, science gives us the facts of the sensory level (the eye of flesh). Above that are the art, morals, and logic of the mental realms (the eye of mind) and, above that, the religion and mysticism of the spiritual realms (the eye of contemplation). Science is relegated to low man on the totem pole, a role that modern science has utterly refused to accept, and in fact will not accept, which is precisely why this traditional epistemological pluralism cannot command the respect—or the cooperation—of modern science.

But in a more sophisticated integration, *each of those levels* (sensory, mental, spiritual) *is also divided according to the differentiations of modernity* (art, morals, and science). Thus—and I must put this very loosely for an introductory statement—there are the art and morals and science of the sensory realm, the art and morals and science of the mental realm, and the art and morals and science of the spiritual realm.

It is exactly this type of synthesis, should it prove sound, that would indeed satisfy *both* the core claim of spirituality (namely, the Great Chain) and the core claim of modernity (namely, the differentiation of the value spheres).

Here science, far from being on the bottom rung, has a hand to play in accessing each of the levels of the Great Chain, from the lowest to the highest (sensory science, mental science, spiritual science). It is not that spirituality takes up where science leaves off, but that they both develop up the Great Chain together. Science is not *under* but *alongside*, and this profoundly reorients the knowledge quest, placing premodernity and modernity hand in hand in the quest for the real, and thus bringing science and religion together in a most intimate embrace.

3

PARADIGMS: A WRONG TURN

Of all the previous attempts to integrate science and religion, by far the most influential and infectious, at least among today's counterculture and a substantial portion of academe, is that of the postmodern/paradigm—the notion that science is actually governed by "paradigms," and a paradigm is simply one of many possible interpretations of reality, no more binding than any other. Since, it is said, paradigms are culturally constructed, not discovered, the authority of science is dramatically undercut, and this leaves room, it is further said, for a "new paradigm" that would be compatible with a spiritual or holistic worldview.

Although the claim is made that this "new-paradigm" approach will at long last integrate scientific and spiritual realities, in fact it totally cripples any effective integration of science and religion. To understand why this is so, we turn to Thomas Kuhn, and to the utterly strange case of his reception by the counterculture.

Although we will devote Part II to an overview of the previous attempts to integrate science and religion—including the postmodern/paradigm approach—it is necessary to address this particular approach right away, before we go any further, and for the simple reason that it has overwhelmingly dominated today's discussion on science and spirituality. So much so that as soon as most people hear about "integrating science and religion," they almost immediately think "new paradigm."

Why that approach is a dead end is the topic of this chapter.

THE MISREADING OF THOMAS KUHN

Thomas Kuhn's *The Structure of Scientific Revolutions* was published in 1962; it soon became, for reasons good and bad, the most influential book on the philosophy of science ever written and the most frequently cited academic book of recent times. For reasons that will become obvious, this book had almost no influence on the *historians* of science; but it dominated discussions on the *philosophy* of science, and—in a great ironic twist—became perhaps the most influential *misunderstood* book of the century. Most of its popularity stemmed from a widespread and massive misunderstanding of its central conclusions, a misunderstanding that, many historians now agree, stemmed in large part from the narcissistic mood of the sixties "Me Generation." How sixties narcissism massively distorted Kuhn is itself a paradigm of our times— and a stunning cautionary tale for anyone stepping into the "science-and-religion" game.

Distortions of Kuhn have now become so common that serious scholars of his work have no trouble reciting the popular misunderstanding of the notion of a "paradigm." Here is Frederick Crews reciting the typical (wrong) view: "Kuhn, we are told, demonstrated that any two would-be paradigms, or regnant major theories, will be incommensurable; that is, they will represent different universes of perception and explanation. Hence no common ground can exist for testing their merits, and one theory will prevail for strictly sociological, never empirical, reasons. The winning theory will be the one that better suits the emergent temper or interests of the hour [ideology, class, prejudice, gender, race, power, etc.—in other words, androcentric, ethnocentric, phallocentric, eurocentric, anthropocentric, and so forth]. It follows that intellectuals who once trembled before the disapproving gaze of positivism can now propose sweeping 'Kuhnian revolution- ary paradigms' of their own, defying whatever disciplinary consensus they find antipathetic. . . ."

That is indeed the typical interpretation of Kuhn. Crews calls that interpretation "theoreticism," because it is a view lost in mere abstract theory divorced from actual *evidence*. He then points out the obvious: "One can gauge the emo- tional force of theoreticism by the remoteness of this inter- pretation from what Kuhn actually said."

The idea was that, since "paradigms" govern science, if you don't like the worldview of science, then simply think up a new paradigm for yourself (this, we will see, is where "narcis- sism" starts to creep into the picture). Since paradigms are al- legedly not anchored in actual facts and evidence (but instead create them), you needn't be tied to the authority of science in any fundamental way. Indeed, science becomes merely one of numerous different readings of the text of the world, with no more actual authority than poetry, astrology, or palmistry: all are equally legitimate interpretations of the blooming buzzing confusion of experience.

This popular (mis)understanding of Kuhn—this "theoreti- cism"—also meant that science was allegedly *arbitrary* (it is the result not of actual evidence but of imposed power struc- tures), *relative* (it reveals nothing that is actually constant in reality but simply things that are relative to the scientific im- position of power), *socially constructed* (it is not a map corre- sponding to any actual reality but a construction based on social conventions), *interpretive* (it does not reveal anything fundamental about reality but is simply one of many inter- pretations of the world text), *power-laden* (it is not grounded in neutral facts; it is not dominated by facts; it simply domi- nates people, usually for ethnocentric and androcentric rea- sons), and *nonprogressive* (since science proceeds by ruptures or breaks, there can be no cumulative progress in any of the sciences).

Kuhn maintained none of those views. Indeed, he vehe- mently argued against most of them. But what Crews so un- erringly called *the emotional force* of the misunderstood idea had already taken root: imagine, we can abandon the strait-jacket of science and evidence by merely thinking up a new paradigm ("merely thinking up" = "theoreticism"; and this itself, as Crews himself points out, was grounded in a rampant sixties narcissism).

A small list of claimants to the "new paradigm" included neoastrology, ecofeminism, deep ecology, altered states of consciousness, the quantum self, the quantum society, systems theory, process philosophy, nonordinary states of consciousness, holistic health, global ecological consciousness, postmodern poststructuralism, quantum psychotherapy, deconstruction, neo-Jungian psychology, channeling, premodern indigenous tribal consciousness, neopaganism, Wicca, palmistry, and the Internet.

Kuhn himself watched all of this with growing alarm, and made a series of vigorous statements meant to curtail the damage, but to absolutely no avail. Most people using the term "paradigm" and citing Kuhn didn't even know that he had abandoned the term. Is science actually relative, arbitrary, and nonprogressive? Kuhn in exasperation: "Later scientific theories are better than earlier ones for solving puzzles in the quite often different environments to which they are applied. This is not a relativist's position, and it displays the sense in which I am a convinced believer in scientific progress." Obviously, we cannot have real scientific *progress* if paradigms are arbitrary, incommensurable, or relative, with none of them intrinsically better than another.

What, then, did Kuhn mean by "paradigm," and what was the "structure" of scientific revolutions? Nothing nearly as dramatic as postmodern theoreticism proclaimed. To begin with, Kuhn outlined not three or four paradigm shifts in the history of modern science, but *several hundred.* As Ian Hacking summarizes the actual view: *The Structure of Scientific Revolutions* "is about hundreds of revolutions, which are supposed to occur in many disciplines, and which typically involve the research work (in the first instance) of at most a hundred or so investigators. Lavoisier's chemical revolution counts as one, but so does Roentgen's discovery of X-rays, the voltaic cell or battery of 1800, the first quantization of

energy, and numerous developments in the history of thermodynamics."

In other words, almost any new experiment generating
new data was a new paradigm, which is why a battery was
a new paradigm. "Paradigm" itself carried two broad components, which we might call "practical" and "social." Kuhn
"used the word [paradigm] to denote both the established
and admired solutions that serve as models of how to practice the science [this is the practical component, a set of exemplars or experiments or injunctions], and also for the local
social structure that keeps those standards in place by teach-
ing, rewards, and the like [the social component, which is
also a set of injunctions or social practices]. The word was
mysteriously launched or rather catapulted into prominence,
and now seems a standard item in the vocabulary of everyone who writes about science—except Kuhn himself, who
has disavowed it. . . . It is at present a dead metaphor."

REAL PARADIGMS

What is not dead—and what Kuhn did not disavow—is that
science is grounded in *injunctions*, exemplars, and social practices. Science is not merely an innocent reflection on a pregiven world, but rather discloses data through injunctions or
exemplars ("exemplar" is a word Kuhn used interchangeably
with "paradigm"). Both components of the paradigm (practical and social) are *grounded in injunctions*, in actual practices,
which is why almost any novel experiment that actually produced new data was viewed as a revolution or "new paradigm." This is why Kuhn counted hundreds of revolutions or
new paradigms, including X-rays and the battery.

But not one of those new paradigms was merely theoretical (that would be "theoreticism"). Rather, they were all
grounded in *evidence* that could be brought forth and reproduced with the given exemplar, paradigm, or injunction. This
is why Kuhn's most common use of "paradigm" was "retool-
ing operations with important consequences for research
practice"—in other words, specific concrete injunctions. And

this is exactly why science can and does show real *progress:* the injunctions or exemplars or paradigms disclose actual evidence, they do not fabricate it based on mere conventions. As Crews notes, "Kuhn happens to be a fervent believer in scientific progress, which, he argues, can occur only after a given specialty has gotten past the stage of what he calls 'theory proliferation' and 'incessant criticism and continual striving for a fresh start.' By incommensurability Kuhn never meant that competing theories are incomparable but only that the choice between them cannot be entirely consigned to the verdict of theory-neutral rules and data. Transitions between paradigms—which in any case are mere *problem solutions, not broad theories* . . . —must indeed be made . . . through 'gestalt switches,' but the rationality of science is not thereby impaired. As Kuhn asked, and as he has continued to insist with mounting astonishment at his irrationalist fan club, 'What better criterion than the decision of the scientific group could there be?' "

To which "new-paradigm" theoreticians of every conceivable sort, astonishingly citing Kuhn, replied, "My new paradigm." This blatant misreading of Kuhn erased evidence from the scene of truth, and into that vacuum rushed every egocentric project imaginable. Science was reduced to rubble, or, more precisely, poetry. As Howard Felperin typically put it— and as one can find echoed in thousands of "new-paradigm," "new age," "transformational" theories—"Science itself is recognizing that its own methods are ultimately no more objective than those of the arts." Science and poetry stand on exactly the same epistemic footing, and this allows us to deconstruct the authority of science right at the start, thus making room for whatever religion we want.

THE POSTMODERN SCENE:
NIHILISM AND NARCISSISM

Once this massive distortion of Kuhn was in place, "new paradigm" thinkers in America began to connect this mis-

Kuhnian notion with every sort of French parlor game, and this unholy mixture of mis-Kuhn and postmodern poststructuralism has come to dominate everything from the new historicism to premodern tribal revivals to postmodern ecophilosophies to "the new holistic paradigm" to cultural studies in general. Since grounding in facts and evidence is no longer required, slogans are treated as facts, as one critic forcefully noted: "In the human studies today, it is widely assumed that the positions declared by . . . poststructuralism are permanently valuable discoveries that require no further interrogation. Thus one frequently comes upon statements of the type: 'Deconstruction has shown us that we can never exit from the play of signifiers'; 'Lacan demonstrates that the unconscious is structured like a language'; 'After Althusser, we all understand that the most ideological stance is the one that tries to fix limits beyond which ideology does not apply'; 'There can be no turning back to naive pre-Foucauldian distinctions between truth and power.' Such servility constitutes an ironic counterpart of positivism—a heaping up, not of factual nuggets, but of movement slogans that are treated as fact."

All of which horrified Kuhn. And yet, as Crews points out, "Nothing Kuhn can say, however, will make a dent in theoreticism, which is less a specific position than a mood of antinomian rebellion and self-indulgence." Time and again Crews hits the notion of self-indulgence and narcissism, and he is by no means alone. He points to "theoreticism, whose purest impulse is toward positing ineluctable constraints on the perceptiveness and adaptability of everyone but the theorist himself." Such self-indulgence, he says, "comes down to us from the later Sixties."

Historian Ernest Gellner, among many others, has made a similar point, namely, that where evidence is erased, narcissism flourishes. The *demand for evidence*—or validity claims—which has always anchored genuine and progressive science, simply means that one's own ego cannot impose on the universe a view of reality that finds no support from the universe itself. The validity claims and evidence are the ways in which we attune ourselves to the Kosmos. The validity

claims force us to confront reality; they curb our egoic fan-
tasies and self-centered ways; they demand evidence from
the rest of the Kosmos; they force us outside of ourselves!
They are the checks and balances in the Kosmic Constitu-
tion.

But it was exactly these checks and balances, these curbs
on narcissism, that the mis-Kuhnian "new-paradigm" thinkers
of almost every variety implicitly or explicitly attempted to
erase. And behind it all lay, in part, the "culture of narcissism." Philosopher David Couzens Hoy points out that "freeing [theory] from its object"—that is, erasing the demand for
evidence—"may open it up to all the possibilities of rich
imaginations; yet if there is now no truth of the matter, then
nothing keeps it from succumbing to the sickness of the
modern imagination's obsessive self-consciousness." Theory
thus becomes "only the critic's own ego-gratification." The
culture of narcissism. "Then a sheer struggle for power en-
sues, and criticism becomes not latent but blatant aggression," part of "the emergent nihilism of recent times."

From the notion that "We are in the midst of a world-
transforming paradigm shift" to the idea that "You create
your own reality"—the many permutations of "self-indulgent
theoreticism": ideas disconnected from the demand for evidence, science reduced to poetry, narcissism and nihilism
joined in a postmodern paradigm from hell.

These critics are not saying that all of the poetry/paradigm
stance is merely or even especially due to narcissism. They
are simply saying—and I agree—that the vaunted narcissism
of the Me Generation *predisposed* many individuals to seize
upon a profound *misreading* of Thomas Kuhn, a misreading
that allowed them arbitrarily to deconstruct any reality that
happened not to suit them and then insert their own "revolutionary new paradigm" into the scene, imagining that they
were somehow vanguards of a revolutionary transformation
that would shake the world to its very foundations, and the
keys to which, they now held.

Thus, as historians puzzle over how a *distortion* of Kuhn
became one of the most cited and most influential notions of
the past three decades, as they puzzle over how an untruth

came to be so wildly influential, they are forced to look for other forces driving this avalanche of error, and the "culture of Narcissus"—whether in the new age, in art criticism, in literary theory, in tribal revivals, in the new historicism, in cultural studies, in me-spirituality, in the idea that you create your own reality, in "the holistic new paradigm"—is appearing the ever more likely candidate, infecting a generation that, subtly but insistently, needed to see itself as central to the unfolding of the universe.

THE PERFORMATIVE
CONTRADICTION

Whether the "new-paradigm" approaches are indeed shot through with narcissistic self-indulgence is an open question. What is not open to question is the fact that these approaches—and extreme postmodernism in general—are internally self-contradictory. They collapse under their own weight, leaving everything from literary theory to the integration of science and religion *in an even worse state* than when they began.

The moment of truth in the postmodern argument is that, indeed, the world is not an innocent perception. The world is in part a construction, an interpretation. This is one of the enduring truths brought forth by postmodernism in general. (Indeed, Chapter 9 is devoted to an appreciation of many of the important insights of the postmodern movement).

But—and here we must part ways with *extreme* postmodernism—all interpretations are not equally valid: there are better and worse interpretations of every text. *Hamlet* is not about a fun family picnic in Yellowstone Park. That is a very bad interpretation, and it can be thoroughly *rejected* by any community of adequate interpreters. All interpretations are not created equal—and that brings to a crashing halt the major claim of extreme postmodernism.

The difficulty is that, in its totalizing attack on truth ("There is no truth, only different interpretations"), extreme postmodernism *cannot itself claim to be true.* Either it must exempt itself from its own claims (the narcissistic move), or what it says about everybody else is equally true for itself, in which case, what it says is not true, either. As Gellner summarizes the disaster: "So, if true, it is false; so, it is false."

This so-called *performative contradiction* in extreme postmodernism has now been pointed out by numerous scholars, including Jürgen Habermas, Charles Taylor, Karl-Otto Apel, Ernest Gellner, among others. Indeed, there is now something of a consensus among serious scholars that extreme postmodernism is a dead end. It either nihilistically denies truth, including therefore its own; or, attempting to avoid that, it retreats into narcissism, exempting itself from its own claims (this is still the popular "new-paradigm" approach).

But the fact remains that the notion of paradigm *"is at present a dead metaphor."* Thus, in our attempt to integrate science and religion, we will have to look elsewhere for the key.

THE SPIRITUAL CRITIQUE OF "NEW PARADIGMS"

Not only have serious scholars, including Kuhn, abandoned the notion of paradigm (as popularly understood); the great wisdom traditions themselves more often than not find the notion utterly confused.

The perennial core of the wisdom traditions is, recall, the Great Chain of Being and the correlative belief in epistemological pluralism. As Huston Smith summarizes this view, "Reality is graded, and with it, cognition." That is, there are levels of both being and knowing. If we picture the Great Chain as composed of four levels (body, mind, soul, and spirit), there are four correlative modes of knowing (sensory, mental, archetypal, and mystical), which I usually shorten to the three eyes of knowing: the eye of flesh (empiricism), the eye of mind (rationalism), and the eye of contemplation (mysticism).

Empirical science, according to epistemological pluralism, can tell us much about the sensory domain and a little bit about the mental domain, but virtually nothing about the contemplative domain. And no "new paradigm" is going to alter that in any way. Chaos theories, complexity theories, systems theories, quantum theories—none of them requires scientists to take up contemplation or meditation in order to understand those "new paradigms," and thus none of them gives any direct spiritual knowledge at all. They are just more mental ideas hooked to sensory perceptions; they are not transmental contemplation disclosing the Divine.

Worse, the wisdom traditions continue, by presenting these new scientific theories as if they were spiritual realities, these "new paradigms" often discourage people from taking up actual contemplation and thus directly accessing Spirit itself. These "new paradigms" in effect replace the eye of contemplation with the eyes of mind and flesh, thus destroying the only mode that is our salvation. Far from helping integrate science and religion, these approaches devastate the true religious impulse.

TO WHOM AM I SPEAKING?

I am in substantial agreement with that criticism, which can also be put in a more modern light, as follows: the eye of flesh is monological; the eye of mind is dialogical; and the eye of contemplation is translogical.

Monological comes from "monologue," which means a single person talking by him- or herself. Most empirical science is monological, because you can investigate, say, a rock without ever having to talk to it. Empirical science carefully chooses objects of research that it will never have to talk to. Whether those objects are rocks, planets, atoms, cells, geological structures, DNA molecules, brain synapses, kidneys, rivers, atmospheric dynamics, ideal gases, thermodynamic bodies, process patterns, systems interactions, ecosystems, it doesn't matter: you don't have to talk to any of them. This is a monological endeavor, tied to the eye of flesh and the data of the human senses or their instrumental extensions.

Dialogical comes from "dialogue," which means talking with somebody and attempting to understand that person. And whereas the eye of flesh is monological, the eye of mind is, in many important ways, dialogical. As you are now reading these sentences, you are involved in a dialogical mode of knowing. You are attempting to understand what I mean by these symbols. If I were actually present, you might ask me directly, and we would talk. We would be involved in interpretation, in hermeneutics, in symbolic meaning, in mutual understanding. You are not treating me as an *object*, like a rock, which you will stare at monologically; you are treating me as a *subject*, which you will try to understand dialogically.

Translogical means transcending the logical, the rational, or the mental in general. Formless mysticism, disclosed with the eye of contemplation, is translogical: it sees beyond the eye of flesh (and its monological empiricism) and beyond the eye of mind (and its dialogical interpretation), and instead stands open to the radiant Divine (in nondual gnosis). This spiritual opening can be directly accessed by neither the eye of flesh nor the eye of mind, only the eye of contemplation. And the very heart of the great wisdom traditions is a contemplative opening to the spiritual domain, which is not monological, not dialogical, but translogical.

Perhaps you can start to see why even the great wisdom traditions (with their epistemological pluralism) can offer such a devastating critique of the notion that a "new paradigm" in science would, or even could, be equivalent to a spiritual opening. For what is required is not a new monological science or a new dialogical interpretation, but a genuine method for directly opening to translogical contemplation, and no "new scientific paradigm" whatsoever has been able to make that offer.

It is common among the "new-paradigm" thinkers to claim that the basic problem with science is that, under the "Newtonian-Cartesian" worldview, the universe is viewed as atomistic, mechanistic, divided, and fragmented, whereas the new sciences (quantum/relativistic and systems/complexity theory) have shown that the world is not a collection of atomistic fragments but an inseparable web of relations. This "web-of-life" view, they claim, is compatible with traditional spiritual worldviews, and thus this "new paradigm" will usher in the new quantum self and quantum society, a holistic and healing worldview disclosed by science itself.

But that approach, according to the great wisdom traditions, completely misses the point. For the real problem with empirical science is *not* that it is atomistic instead of holistic, or that it is Newtonian instead of Einsteinian, or that it is individualistic instead of systems-oriented. The real problem is that *all* of those approaches—atomistic and holistic alike—are monological. They are all empirical and sensorimotor based—evidence supplied by the senses or their instrumental extensions. This is true of Newtonian science and of Einsteinian science. It is true of atomistic science and of systems science. Under no circumstances—and under no paradigm whatsoever—does empirical science show any inclination to deny its empiricism—nor should it.

No, the real problem of our modern fragmentation is not that empirical science is atomistic rather than systems-oriented; the real problem is that *all higher modes of knowing have been brutally collapsed into monological and empirical science.* Both atomism and systems theory are monological/empirical, and *it is the reduction of all knowledge to monological modes that constitutes the disaster of modernity.* The higher modes themselves—mental and supramental, rational and transrational, hermeneutic and translogical, contemplative and spiritual—have all been rudely reduced and utterly collapsed to the eye of flesh and its extensions, and whether that monological madness be atomistic or systems-oriented is quite beside the point.

Has there been a recent revolution in science, a genuine *new paradigm in science itself* that is holistic rather than atomistic? Yes, definitely. There have been several of them, actually, including various aspects of quantum physics, relativistic physics, cybernetics, dynamical systems theory, autopoiesis, chaos theory, and complexity theory. These are all new revolutions, new paradigms in the true sense, with new modes of research, new social practices supporting them, new types of data, new forms of evidence, and new theories surrounding them.

But they are all, without exception, monological to the core. And thus, as important as they are in their own right, they have little to offer us in terms of actually integrating monological with dialogical and translogical—that is, integrating science and spirituality.

So we can begin to see that, although it is common for "new-paradigm" thinkers to claim that their scientific systems orientation will heal the fragmentation of the modern world, make us feel at home in the universe, save the planet, and reintroduce spirituality into our sick and alienated culture, the fact is that monological science—both atomistic and systems-oriented—is, alas, part of the disease it claims to cure.

SUMMARY

It is not through any sort of "new paradigm" in science that spirituality and modern science will finally find mutual accord. Because, first, "paradigm" as popularly understood "is at present a dead metaphor." Second, even if it were alive—or used in its correct meaning as injunction or social practice—there is still nothing in even the most avant-garde of the empirical sciences (from string theory to hyperspace to chaos theory) that goes beyond its monological/empirical grounding. All of the alleged "new paradigms" would still appear within that monological framework, not outside it, and thus any and all of the "new paradigms" would simply clone the disaster. Third, the entire approach, according to many critics, is heavily infected with a narcissistic disregard for evidence. Fourth, the approach is profoundly self-contradictory (the performative contradiction). Fifth—and worst of all—it is based on a category error, the attempt by the monological eye of flesh and the dialogical eye of mind to see what can be seen only by the translogical eye of spirit. As such, it can profoundly detract from the awakening of a genuine spiritual awareness.

What is required for an integration of science and religion is not an attempt to reduce translogical religion to a new monological paradigm. Rather, we need to take the core of the wisdom traditions—namely, the Great Chain of Being, which includes monological (the eye of flesh) and dialogical (the eye of mind) and translogical (the eye of contemplation)—and expose them to the differentiations of modernity (the differentiation of the value spheres of art, morals, and science).

To do that, we need to understand as clearly as we can the beast called "modernity."

4

MODERNITY: DIGNITY AND DISASTER

The great difficulty with all of the typical attempts to integrate premodern religion with modern science is, I believe, a
failure to grasp either the core of premodernity (the Great
Chain) or the core of modernity (the differentiation of the
value spheres of art, morals, and science). Since there seems
to be less confusion about the core of premodernity—the
Great Nest of Being was its heart and soul—perhaps we need
to look more carefully at the other side of the equation, the
monster known as "modernity."

THE MEANING OF "MODERNITY" AND "POSTMODERNITY"

Modernity, for historians, refers very loosely to the general
period that had its roots in the Renaissance, blossomed with
the Enlightenment, and continues in many ways to this day.
Modernity therefore includes various trends in:

Philosophy: Descartes is considered the first "modern"
philosopher; modern philosophy is usually "representational," which means it tries to form a correct representation of the world. This representational view is also called
"the mirror of nature," because it was commonly believed

that the ultimate reality was sensory nature and philosophy's job was to picture or mirror this reality correctly.

Art: Modern art in the most general sense (from the middle
of the eighteenth century forward)—that of Goya, Consta-
ble, Courbet, Manet, Monet, Cézanne, van Gogh, Matisse,
Kandinsky—is marked at times by an almost total break
with traditional themes and modes of composition, and especially a break from depicting merely mythic-religious
themes (nature, not myth, comes more to the fore).

Science: Modern science (Kepler, Galileo, Newton, Kelvin,
Watt, Faraday, Maxwell) relied in large part on the measurement of empirical-sensory data. The old sciences had
classified nature, the new sciences *measured* nature; and
that was their astonishing and revolutionary strength.

Cultural cognition: This involved a general shift from
mythic-membership modes of cognition to mental-rational
modes; a shift from conventional to postconventional
ethics; a shift from ethnocentric values to universal or
global values.

Personal identity: This involved a shift from a role identity,
defined by a social hierarchy, to an ego identity, defined by
personal autonomy.

Political and civil rights: This included the outlawing of slavery, the institution of women's rights, child labor laws, the
rights of humankind (freedom of speech, religion, assembly, fair trial), and equality before the law.

Technology: This refers especially to inventions beginning
with the steam engine, as well as industrialization in general.

Politics: This included the rise of the liberal democracies,
often through a series of actual revolutions (in, e.g., France
and America).

Will and Ariel Durants' description of modernity as the
"Age of Reason and Revolution" is as good a summary as any.

While historians basically agree on the general outlines of
modernity, *postmodernity* has an extraordinary number of
meanings, few of which coincide. "Postmodern" is often given
both a narrow or technical meaning, and a broader and more
general meaning. The narrow and technical we discussed
briefly in the previous chapter—the notion that there is no
truth, only interpretation, and all interpretations are socially

constructed. This narrow view we also called "extreme postmodernism," because it takes certain very important insights (e.g., many realities are socially constructed) and blows them totally out of proportion (e.g., all realities are socially constructed), which results in nothing but severe performative contradictions.

But in the broader and more general sense, "postmodern" simply means any of the major currents occurring *in the wake of modernity*—as a reaction against modernity, or as counterbalance to modernity, or sometimes as a continuation of modernity by other means. Thus, if industrialization is modern, the information age is postmodern. If Descartes is modern, Derrida is postmodern. If perspectival rationality is modern, aperspectival network-logic is postmodern. If Bauhaus architecture is modern, Frank Gehry is postmodern. If representation is modern, nonrepresentation is postmodern. If the internal combustion engine is modern, the Internet is postmodern. (I will use both of those meanings—the narrow and the more general—as the context will make clear.)

Thus, today's "modern world" actually consists of several different currents, some of which are "modern" in the specific sense (those events set into motion with the Western Enlightenment, as listed above), others of which are carryovers from the premodern world (in particular, remnants of mythic religion, and, more rarely, remnants of tribal magic), and still others of which are postmodern. In short, today's "modern world" actually consists of various premodern, modern, and postmodern currents.

When I refer to *modernity* per se, I mean modernity in the specific sense (the events set into motion with the liberal Enlightenment), whereas "the modern world" simply means today's contemporary world with all of its premodern, modern, and postmodern currents. And it is especially modernity in the specific sense that we wish to understand, because the core claims of modernity must obviously be an essential feature of any genuine integration of modern science and premodern religion.

And, most significantly, the dramatic failure to grasp the actual contours of modernity has crippled more than one attempt to integrate science and spirituality.

THE DIGNITY OF MODERNITY

In many ways, the *governing principles* of the hundred or so democratic nations in today's world are in fact the *principles of modernity*—that is, the values of the liberal Western Enlightenment. These include the values of equality, freedom, and justice; representational and deliberative democracy; the equality of all citizens before the law, regardless of race, sex, or creed; political and civil rights (freedom of speech, religion, assembly, fair trial, etc.). Of course, some of these rights still need to be applied more universally and evenhandedly, but they are nonetheless firmly ensconced as widely held ideals toward which liberal societies ought to strive.

These values and rights existed nowhere in the premodern world on a large scale, and thus these rights have been quite accurately referred to as the *dignity of modernity.* For example, as Gerhard Lenski has documented, every one of the premodern societal types—including tribal, foraging, horticultural, and agrarian—had various degrees of slavery. The only societal types in all of history and prehistory to effectively ban slavery across the board were those that emerged in the wake of modernity. This is simply one example of the many dignities the Enlightenment brought.

To say that no premodern societal type possessed these various dignities is also to say, most damningly, that none of the premodern religions anywhere in the world delivered these dignities and rights on any sort of large scale; in fact, they often did quite the contrary. The battle cry of the Enlightenment—Voltaire's "Remember the cruelties!"—was a call to end the brutal oppression often effected by premodern religion in the name of a chosen God or Goddess. The temples to those Deities were built on the broken backs of millions, who left a trail of blood and tears on the highway to that heaven.

The fact that premodern religion failed to deliver these dignities serves as a sharp reminder that "Godless modernity" was not merely the monster its religious opponents have often claimed. Modernity brought these dignities, so it is to modernity that we will want to look for those factors that supported them. Whatever it was that allowed modernity to bring forth these noble values will be a necessary ingredient in the integration of the best that both epochs have to offer.

MODERNITY AND ITS
LEGION OF CRITICS

Almost all of the "new-paradigm" thinkers couple their proposed new paradigm with an aggressive attack on modernity, resorting at times to vicious polemic. Toxic treatise after toxic treatise tears into modernity, with such typically titled books as *My Name is Chellis and I'm in Recovery from Western Civilization* (actual title). Yet, almost without exception, these "new-paradigm" thinkers give little indication that they have grasped or understood the actual nature of modernity itself—its defining characteristics, values, and structures. In particular, they rarely evidence a clear and concise understanding of the dignity of modernity, even though they implicitly and extensively exercise it.

Instead, they often set up a truly pitiful straw man, often centered on poor Newton and Descartes, and then proceed to damn virtually the entire sweep of modernity. The "Newtonian-Cartesian patriarchal alienated fragmented worldview"—which, of course, is now labeled the "old paradigm"—will be replaced with the revolutionary and world-transforming "new paradigm," which these theorists possess and are willing to share with the world in preparation for the coming transformation.

The various "new paradigms" these theorists offer generally fall into one of three broad types (although combinations are common): premodern revivalist, postmodern pandemonium, and global systems. While all of them possess important moments of truth, which need to be fully acknowledged, virtually all of them fail lamentably in their overall grasp of modernity.

The *premodern revivalist* "paradigm" generally maintains that tribal foraging cultures possessed "nondissociated consciousness," whereas the modern world has mostly "dissociated" or "fragmented" consciousness. Alternatively, the premodern world is viewed as matrifocal and holistic, plugged into the Goddess and the unbroken Web of Life, whereas the modern world is patriarchal, analytic, fragmented, and broken. Thus what the modern world needs is a *resurrection* or a *recapturing* of a lost and more "unified" consciousness. But, as we will see, these writers tend to drastically misinterpret the premodern consciousness—it was far from "unified" in most instances. Moreover, what none of these theories has been able to satisfactorily explain is why evolution would do something that it has never done in any other living system, namely, make a U-turn right in the middle of its development, rather like every oak tree on the face of the planet suddenly attempting to recapture its acornness.

The *postmodern* "paradigm" ("postmodern" in the narrow and technical sense) is simply the claim that there is no truth, only interpretations, and thus the "sliding nature of all signifiers" means that the authority of science—and therefore modernity itself—can simply be swept under the carpet with not much further ado. We are free of modernity precisely because we are free of the demand for truth and verification in general. The very demand for truth is part of the "old paradigm," which the new paradigm has totally deconstructed. This leaves, as we saw, nothing but one's own ego—one's own narcissism—to impose its will on reality, and this nihilistic narcissism is boldly offered to the world as a revolutionary transformation.

The *global systems* "paradigm" attacks atomism and replaces it with systems thinking, imagining that it has thereby bypassed the central problem of the "Newtonian-Cartesian fragmented worldview." But, as we saw, the specific difficulty with empirical science of *any* variety is not that it is atomistic or holistic, analytic or systems, but rather that it is empirical and monological in the first place. Systems theory does not alter that in the least; it merely continues the monological madness by other means, which, in this case, is all the more insidious because its proponents imagine that they have overcome the problem, whereas they have simply cloned it.

The grave difficulty with all three of these attacks on modernity—other than their own performative contradictions—is that few of them evidence any substantial understanding of the characteristics, let alone the dignity, of modernity. Ironically, most of the decent values that these approaches express are in fact the values of modernity, including equality, freedom, justice, equal opportunity, and egalitarianism before the law. This certainly gives the impression of ungrateful and petulant children not on speaking terms with their parents.

Most egregious, these "new-paradigm" attacks on modernity show no evidence of understanding the difference between *differentiation* and *dissociation*. Yet in that simple but profound distinction lies the key to modernity—and therefore the key to the integration of science and religion in the modern world.

DIFFERENTIATION = DIGNITY

Scholars from Max Weber to Jürgen Habermas have sought a simple way of characterizing the major thrust that was modernity, and they have hit upon what is surely one of the most significant developments in all of human history: as we have briefly seen, modernity was characterized by what Weber called "the differentiation of the cultural value spheres"—that is, *the differentiation of art, morals, and science.* This differentiation is the essence of the dignity of modernity, as a brief glance at premodernity will show.

Many scholars (including Jean Gebser, Habermas, myself, and others) divide the premodern world into archaic, magic, and mythic worldviews (correlated with foraging, horticultural, and agrarian modes of production; these terms will become clearer as we proceed). But none of the premodern worldviews clearly differentiated art-aesthetics, empirical-science, and religion-morals. Although "premodern holists" claim that this is a wonderful state of nondissociated and unified consciousness, it is actually quite the opposite.

The Church during the Middle Ages is a classic example, repeated around the world and in every premodern societal type as variations on a common theme. Because art-aesthetics, empirical-science, and religion-morals were not clearly differentiated, what happened in one sphere could dominate and control what happened in the others. Thus, a scientist such as Galileo could be prevented from pursuing the sphere of science because it clashed with the prevailing sphere of religion-morals. An artist such as Michelangelo was in constant conflict with Pope Julius II about the types of figures he was allowed to represent in his art, because expressive-art and religion-morals were not clearly differentiated, and thus oppression in one sphere was oppression in the other.

Likewise, the state was not yet differentiated from religion—there was no separation of church and state. Accordingly, if you disagreed with the religious authorities, you could be tried for both heresy (a *religious* crime) and treason (a *political* crime). For heresy, you could be eternally damned; for treason, temporally tortured and killed—and those who committed the former usually suffered the latter. Few of the theorists who glowingly eulogize the numerous premodern theocracies (or mythocracies) as "organic and unified" would actually want to live in such a culture, because if your religion did not happen to agree with that of the authorities, you were toast.

This state of affairs was not holistic and integrated, it was simply *predifferentiated*—a huge contrast! That which has not yet been differentiated in the first place cannot be integrated. There were as yet no separate spheres to be brought together into a synthesis or integration; there was simply a fusion of spheres that robbed each of its autonomy and dignity.

But with the rise of modernity, the spheres of art, science, and morals were clearly differentiated, and this marked the *dignity* of modernity because each sphere could now pursue its own truth without violence and domination from the others. You could look through Galileo's telescope without being hauled before the Inquisition. You could paint the human body in a natural setting without being tried for heresy against God and Pope. You could espouse the universal moral rights of humans without being charged with treason against King or Queen.

So here is our first important equation defining modernity: differentiation = dignity. If we are to integrate modern science with premodern religion, that is one of the essential gifts that modernity will bring.

THE GOOD, THE TRUE, AND THE BEAUTIFUL

There is a simple way of referring to these three value spheres of morals, science, and art: they are the Good, the True, and the Beautiful. (These terms were first introduced on a large scale by the Greeks, who were, in this regard, one of the precursors of modernity.)

The Good refers to morals, to justness, to ethics, to how you and I interact in a fair and decent fashion, both with each other and with all other sentient beings. This does not mean that everybody has to agree on a specific type of morality, about which there can be reasonable disagreement. It means, in a general sense, that human beings must discover some way to mutually inhabit the same cultural space, the opposite of which is, quite simply, war.

The True refers, in a very general sense, to objective truth. It means the truth according to dispassionate standards—not merely the truth according to my ego, or my tribe, or my religion. Science, above all, attempts to specialize in objective, empirical, reproducible truth. This does not mean there are not other types of truth; it simply means that science has a deserved reputation for delivering important types of objective truth.

Beauty, it is said, is in the eye of the beholder; it represents the aesthetic and expressive currents of each subjective self. This does not necessarily mean that beauty is "merely subjective" or idiosyncratic; it simply means that beauty is a judgment made by each subject, each "I." This judgment, as Kant pointed out, resides not empirically in an object, but in a discriminating subject. Beauty is (in part) in the "I" of the beholder.

So to say that modernity differentiated morals, science, and art is to say that it differentiated what is good, what is true, and what is beautiful, so that each of the spheres could pursue its own truths and aspirations without domination or violence from the others.

I, WE, AND IT

We now reach a fascinating and important point in this discussion. Each of these spheres—art, morals, and science; or the Beautiful, the Good, and the True—*has a different type of language.* The expressive-aesthetic sphere is described in "I" language. The moral-ethical sphere is described in "we" language. And the objective-science sphere is described in "it" language. *If we are to integrate these various spheres, we must first learn to speak their native tongues.*

Beauty, we just saw, is in the "I" of the beholder. This *subjective* domain represents the self and self-expression, aesthetic judgment, and artistic expression in the most general sense. It also represents the irreducible subjective contents of immediate consciousness (and intentionality), all of which can properly be described in first-person accounts, in "I" language.

Ethics is described in "we" language. It is part of the *intersubjective* domain, the domain of collective interaction and social awareness, the domain of justness, goodness, reciprocity, and mutual understanding, all of which are described in "we" language.

Truth, in the sense of objective truth, is described in "it" language. This is the domain of *objective* realities, realities that can be seen in an empirical and monological fashion, from atoms to brains, from cells to ecosystems, from rocks to solar systems, all of which are described in "it" language.

Thus, when we say that modernity differentiated the spheres of art, morals, and science, this also basically means that modernity differentiated the realms of I, WE, and IT.

Because modernity differentiated the WE and the IT, political or religious tyranny (of the WE) could no longer determine what was objectively true (of the IT). In other words, you could now read Copernicus without being burned at the stake. This differentiation of WE and IT led directly to the rise of the empirical sciences, including the ecological sciences, systems theory, and quantum-relativistic physics. Premodern societies produced none of these empirical sciences, in part because they lacked this crucial differentiation.

Because modernity differentiated the I's and the WE, the collective WE could no longer dominate individual I's. That is, each individual I had *rights* that could not be violated by the state, the Church, or the community in general. This differentiation of I and WE contributed directly to the rise of the liberal democracies, where each I was extended the political rights of equality, freedom, and justice. This in turn led to such liberation movements as the abolition of slavery, women's rights, and the freeing of the untouchables.

Because modernity differentiated the I and the IT, individual whim could no longer establish what was objectively true. What the I believed about objective reality now had to be checked against empirical facts, thus curbing the magical and mythical attempts to coerce the Kosmos through egocentric ritual and petition. This contributed directly to everything from the rise of modern medical science to global telecommunications—in other words, if I want something from reality (it), I am going to have to do something other than merely wish, since the two are not the same. (This also shows us the disasters that occur when both premodern revivalists and postmodern deconstructionists attempt to dedifferentiate these realms, thus equating I-art with it-science and immersing themselves in exactly the narcissism that this differentiation overcame.)

And so goes the list of differentiated dignities. Liberal

democracy, equality, freedom, feminism, the ecological sciences, the abolition of slavery, extraordinary medical advances, modern physics—all of these rest, in whole or part, on the differentiation of expressive-aesthetics, legal-morals, and empirical-science; the Beautiful, the Good, and the True; I, WE, and IT; self, culture, and nature.

And that is precisely why this series of crucial differentiations is called the "dignity of modernity."

DIFFERENTIATION AND DISSOCIATION

It is this dignity of differentiation that the antimodernity critics so often completely miss. They do so, I believe, because they invariably confuse *differentiation* with *dissociation*.

All natural and healthy growth processes proceed by differentiation-and-integration. The clearest example of this is the growth of a complex organism from a single-celled egg: the zygote divides into two cells, then four, then eight, then sixteen, then thirty-two . . . into literally millions of cells. And while this extraordinary *differentiation* is occurring, the different cells are simultaneously being *integrated* into coherent tissues and systems in the overall organism. This differentiation-and-integration process allows a single cell to evolve into a multicellular organism and complex system of exquisite unity and functional integrity.

From a simple acorn to the mighty oak: the extraordinary process of differentiation-and-integration. In this growth process, if something goes wrong with either of those strands of growth—differentiation or integration—the result is *pathology.*

If differentiation fails to occur, the result is *fusion,* fixation, and arrest in general. Growth becomes stuck at a particular stage; there is no further growth because further differentiation fails to occur. To give an example from human psychosexual growth, when we say some people have an oral fixation, it means that they are fixated to an oral impulse that they failed to differentiate from. They remain "fused" to this impulse, which obsessively dominates their awareness.

On the other hand, if differentiation begins but goes too far, the result is *dissociation* or fragmentation. Differentiation gets out of control, and the various subsystems cannot be easily integrated: they fly apart instead of fitting together. The parts don't differentiate, they dissociate, and the result is fragmentation, repression, alienation.

In human growth, for example, the ego and id are supposed to differentiate; but if that differentiation goes too far into dissociation, the ego simply represses and alienates the id, which results in painful neurotic symptoms. Instead of differentiation and integration, there are dissociation and repression.

Now, if we confuse differentiation with dissociation, we will confuse growth with disease. We will confuse dignity with disaster. We will confuse evolution with catastrophe. And that is precisely what so many of the antimodernist critics do.

Of course, differentiation can indeed look like a split, a separation, a breaking, or a fracture. The one-celled egg does indeed divide into two cells, then four, and so on. But that multiplication is how nature creates *higher unities and deeper integrations*. It is easy to unify a hundred items, but try unifying a million. That is exactly why the unity of the oak is infinitely more impressive than that of the acorn: the oak has considerably more *depth* (it has a much greater number of systems that must be vertically integrated in order to function).

Thus, the prior acorn state is not "more unified," it is simply *less differentiated*—and actually, therefore, considerably *less integrated*. The oak is incomparably more unified and integrated than the acorn, and it got that way precisely through the developmental and evolutionary process of differentiation-and-integration.

But the premodern revivalists, looking at the course of humanity's necessary differentiations (on the way to higher integrations), see nothing but a series of fractures, breaks, dissociations, and disasters. When humanity differentiated mind and nature (around the tenth millennium B.C.E.), the premodernists scream dissociation. When humanity differentiated mind and body (around the sixth century B.C.E.), the premodernists scream dissociation. When humanity finally differentiated art and science and morals (with modernity), the premodernists scream dissociation.

And they see nothing but a sinister, malevolent, or even evil agency behind these brutal "ruptures." The oak is somehow a vicious and horrible violation of the acorn. The exact nature of the evil and disruptive agency responsible for humanity's "fragmentation" varies from theorist to theorist. Top contenders include Newton, Descartes, the patriarchy, Plato, analytic reason, farming, cooking, domestication of animals, mathematics, males in general, belief in a transcendent God, and processed foods.

And, for the premodern revivalists, the cure is somehow to recontact and resurrect our acornness. We must get back to a state prior to the "dissociation." But because these theorists tend to confuse differentiation and dissociation, they confuse dignity and disaster, they confuse forward and backward. They would have us heal the dissociations of modernity, which is well and good; but because they do not distinguish between differentiation and dissociation, they keep looking for a previous period in history where there were *no differentiations at all;* this forces them to look further and further back into prehistory, searching for that state of perfect acornness prior to any nasty divisions. They inevitably end up at one of the earliest stages of human evolution—foraging or horticulture—and this simple state of fusion and indissociation is eulogized as being very close to a state of perfect harmony among mind, body, and nature—when in fact those systems were not integrated, they were simply not yet clearly differentiated in the first place.

Thus the recommendations of these theorists often result in nothing but a thinly disguised *regression.* Of course, none of these theorists actually recommends regression, and the stated idea is always to somehow integrate acornness with oakness (whatever that might mean). But precisely because they show so little evidence of understanding the dignity of modernity, they show little insight into the disaster of modernity, either.

Let us see if we can more accurately spot the specific diseases that did indeed plague modernity. This is crucial, because if we are to integrate premodern religion with modern science, we must know what part of modernity was growth, and what part was disease.

DISSOCIATION = DISASTER

Modernity obviously has its own share of horrible problems. In fact, some of modernity's differentiations did indeed go too far, into a specific set of *dissociations*—and those dissociations I refer to as *the disaster of modernity*. Not only did art, morals, and science differentiate—which was necessary and beneficial—they soon began dramatically to dissociate or fly apart, which, as we just saw, is the hallmark of *pathology* in any growing system.

This was indeed a disaster, a pathology, for it very soon allowed a powerful monological science to colonialize and dominate the other spheres (the aesthetic-expressive and the religious-moral), mostly by denying them any real existence at all! If differentiation was the dignity of modernity, dissociation was the disaster.

This dissociation of the cultural value spheres is exactly what began to happen to art, science, and morals. If the modern differentiation began in earnest around the sixteenth and seventeenth centuries, by the end of the eighteenth and the beginning of the nineteenth the differentiation was already drifting into a painful and pathological dissociation. Art and science and morals began going their separate ways, with little or no discourse between these spheres, and this set the stage for a dramatic, triumphant, and altogether frightening invasion of the other spheres by an explosive science. Within a mere century, *monological science*—variously including positivism, empiric-analytic reason, dynamic process theory,

systems theory, chaos theory, complexity theory, and techno-
logical modes of knowing—would completely dominate serious discourse in the Western world.

Put bluntly, the I and the WE were colonialized by the IT.
The Good and the Beautiful were overtaken by a growth in
monological Truth that, otherwise admirable, became
grandiose in its own conceit and cancerous in its relations to
others. Full of itself and flush with stunning victories, empirical science became *scientism*, the belief that there is no reality save that revealed by science, and no truth save that
which science delivers. The subjective and interior domains—the I and the WE—were flattened into objective, exterior, empirical processes, either atomistic or systems.
Consciousness itself, and the mind and heart and soul of humankind, could not be seen with a microscope, a telescope, a
cloud chamber, a photographic plate, and so all were pronounced epiphenomenal at best, illusory at worst.

The entire interior dimensions—of morals, artistic expression, introspection, spirituality, contemplative awareness,
meaning and value and intentionality—were dismissed by
monological science because none of them could be registered by the eye of flesh or empirical instruments. Art and
morals and contemplation and spirit were all demolished by
the scientific bull in the china shop of consciousness. And
there was the disaster of modernity.

FLATLAND

We can also call this disaster "the collapse of the Kosmos,"
because the three great domains—art, science, and morals—
after their heroic differentiation, were rudely collapsed into
only one "real" domain, that of empirical and monological
science, a world of nothing but meaningless ITs roaming a
one-dimensional flatland. The scientific worldview was of a
universe composed entirely of objective processes, all described not in I-language or we-language, but merely in it-
language, with no consciousness, no interiors, no values, no
meaning, no depth and no Divinity.

And, contrary to what some "new-paradigm" thinkers claim, the worldview of science was, almost from the beginning, a systems or holistic view. The Enlightenment philosophers and scientists conceived of nature and humans as one great, interwoven system, with every aspect perfectly intermeshing with every other. This "great interlocking order," as numerous theorists from Charles Taylor to Arthur Lovejoy have carefully demonstrated, was one of the defining conceptions of the Enlightenment and of the modern scientific worldview.

The problem, in other words, was not that the scientific worldview was atomistic instead of holistic, because it was basically and generally holistic from the start. No, the problem was that it was a thoroughly *flatland holism*. It was not a holism that actually included all of the interior realms of the I and the WE (including the eye of contemplation). It was rather a holism, a systems theory, that included nothing but ITs, nothing but objectifiable processes scurrying through information loops, or gravity acting at a distance on objects, or chemical interactions of atomic events, or objective systems interacting with other objective systems, or cybernetic feedback loops, or digital bits running through neuronal circuits. Nowhere in systems theory (or in flatland holism) could you find anything resembling beauty, poetry, value, desire, love, honor, compassion, charity, God or the Goddess, Eros or Agape, moral wisdom, or artistic expression.

In other words, all you found was a holistic system of interwoven ITs. And it was the reduction of all of the value spheres to monological ITs perceived by the eye of flesh that, more than anything else, constituted the disaster of modernity.

THE FACE OF TODAY

It is true that no premodern cultures had this shuddering dissociation and collapse—but only because none had the differentiation of which this dissociation was a pathology. Premodern cultures did not have this disaster precisely because they did not possess the corresponding dignities, either, and thus they cannot serve as role models for the desired integration. The cure for the disaster of modernity is to address the dissociation, not attempt to erase the differentiation!

This dissociation, this disease, this developmental pathology—this collapse of the Kosmos—is of profound significance in attempting to understand what happened to Spirit in the modern world. The Great Chain of Being—the backbone of every human culture prior to modernity—collapsed in the face of unrepentant ITs. All of the higher levels and spheres, including mind and soul, spirit and goodness and beauty, were meticulously scrubbed from the face of the Kosmos, leaving dirt and dust, systems and sand, matter and mass, objects and its. A cold and uncaring wind, monological in its method and calculated in its madness, blew across a flat and faded landscape, the landscape that now contains, as tiny specks in the corner, the faces of you and me.

THE FOUR CORNERS OF
THE KNOWN UNIVERSE

Even if we thoroughly acknowledge the dignity of modernity, we still must address the disaster of modernity. As we
saw in the last chapter, the nightmare is not that science is
atomistic instead of holistic; the disaster is that science per
se—empirical, monological, instrumental, it-language science, in both its atomistic and holistic forms—came to aggressively invade the other value spheres—including interior
consciousness, psyche, soul, spirit, value, morals, ethics, and
art—thus reducing the entire lot to a colony of science,
which itself would pronounce on what was, and what was
not, real.

What was real was any objectifiable entity or process that
could be described in valueless, empirical, monological,
process it-language. These objects, or ITs, all have what
Whitehead called *simple location:* you can actually or figura-
tively put your finger on them, you can see them with your
senses (or their extensions). Molecules are real, organisms are
real, the brain is real, planets are real, galaxies are real,
ecosystems are real. They are all objective, empirical, exterior, positivistic entities: you can put your finger right on
them, more or less.

But you cannot put your finger on compassion; it does not
have simple location. You cannot put your finger on consciousness; it does not have simple location. You cannot put
your finger on honor, valor, love, mercy, justice, morals, vision,

or satori—none of the *interior* dimensions of the I and the WE have simple location. They are located in interior spaces, not in exterior spaces. You cannot put your finger on them.

And you certainly cannot put your finger on God. God will not be fingered. Therefore, according to science and its belief in simple location, God does not exist. In fact, according to flatland, none of the interior dimensions and modes of knowing has any substantial reality at all. Only objective ITS are real.

The disaster of modernity, in short, was that all *interior* dimensions (of I and WE) were reduced to *exterior* surfaces (of objective ITS), which, of course, completely destroys the interior dimensions in their own terms. With this collapse of the Kosmos, there is no longer serious room for any interior apprehension whatsoever, and whether that interior vision be of poetry or of God matters not in the least: none of them has any substantial or irreducible reality.

This is why modern science is not impressed with epistemological pluralism. What might make sense to enlightened types like you and me—namely, that other modes of knowing disclose other and equally valid realities, so that science and religion can peacefully coexist—is soundly rejected by science at the very start, simply because the desired integration involves terms that science does not believe are real to begin with. Why, science asks, should we attempt to integrate Santa Claus? Why integrate pathology, illusion, and error? Why a holistic inclusion of nonsense?

Thus we confront what is by far the most important and central issue in the relation of science and spirituality, namely, the actual relation of any *interior realities* to *exterior realities*. When modern empirical science rejected the reality of the interior domains, it in effect *rejected the entire Great Chain of Being*, because all of the levels of the Great Chain except the lowest (the material body) happen to be *interior realities* of the I and the WE, of the subjective and the intersubjective domains. To reject the interiors was to reject the Great Chain, and thus profoundly reject the core of the great spiritual traditions.

We can therefore summarize the entire collapse of the Kosmos—and modernity's rejection of the Great Chain—by saying that all interiors were reduced to exteriors. All subjects were reduced to objects; all depth was reduced to surfaces; all I's and all WE's were reduced to ITs; all quality was reduced to quantity; levels of significance were reduced to levels of size; value was reduced to veneer; all translogical and dialogical were reduced to monological. Gone the eye of contemplation and gone the eye of mind—only data from the eye of the flesh would be accorded primary reality, because only sensory data possessed simple location, here in the desolate world of monochrome flatland.

INTERIOR AND EXTERIOR

Here is the central problem, the major reason modern science rejected religion, and the major reason higher and interior modes were replaced by an exterior and monological monopoly:

In the traditional view of epistemological pluralism, and in the traditional Great Chain of Being—let us use the simple version of body, mind, soul, and spirit—the material body, the lowest rung, was available to science, and it was science's role, limited but important, to investigate this material realm. But the mind, soul, and spirit "transcended" the body, and thus had no major referents in the body itself. In some versions of the Great Chain, the higher levels had no connection to the body whatsoever—and thus, it was maintained, science had nothing to contribute to, or even say about, these higher and more significant realities.

But as modern science (freed from its slavery to religious tenets by the differentiation of the value spheres) began investigating the organic body—the organism itself—it found that many of the "higher" or "transcendent" realities were actually deeply connected to the organic body and its organic brain: they were functions of the overall organism, not functions of some pie-in-the-sky realms. Consciousness, for example, seemed intimately connected with the organic brain, a profound recognition that had been completely lacking in virtually every premodern culture.

Thus, if the higher realm of the "transcendent mind" was actually a function of the organic body (or the overall organism)—and if religion and metaphysics had completely missed this elemental connection—why should any of the other "transcendent realms" be any different? Couldn't all of this "metaphysical otherworldliness" actually involve *functions of the natural organism*, the *this-worldly* organism, best investigated by empirical science, and not relegated to invisible realms manipulated by dubious mystics?

When science discovered that mind and consciousness were anchored in the natural organism—and not merely floating around somewhere in "higher" realms—the Great Chain of Being took a colossal hit from which it never recovered. And unless this hit can be addressed in a straightforward fashion, satisfying the essence of *both* the religious and the scientific claims, the chance of any integration between them is slim indeed.

We have seen that in order to integrate science and religion, we need first to integrate the Great Chain with the claims of modernity. We can now see that a large part of this task is to investigate the relation of *interior* and *exterior* realities. The premodern religions gave a great deal of emphasis to the interior modes of knowing (mental and spiritual), whereas modernity, in both its dignity and disaster, gave an unprecedented emphasis to the exterior modes; with scientific materialism, the exterior *alone* was real. In many ways, then, the battle between premodern and modern is a battle between interior and exterior.

It is my main contention that unless we can find a way for *both* of those claims to be true—the transcendental and the empirical, the interior and the exterior—we will never genuinely integrate science and religion.

THE FOUR QUADRANTS

The Great Chain of Being is, of course, a hierarchy: each higher level transcends but includes its predecessors. As we saw in Chapter 1, it is best pictured not as a ladder but as a series of concentric circles or nests, with each wider nest enveloping or enfolding its juniors. In Plotinus's version, for example, there are matter, life, sensation, perception, impulse, images, concepts, logical faculty, creative reason, world soul, nous, and the One, with each higher *development* being an *envelopment* of its predecessors.

Modern systems science likewise has its own general hierarchy, each term of which also transcends and includes its predecessors: subatomic particles, atoms, molecules, cells, tissue systems, organisms, societies of organisms, biosphere, universe.

It is fascinating that *both premodern religion and modern science have a defining hierarchy,* and both of them are composed of *enveloping nests of increasing embrace* (development that is envelopment). And yet, these two major and extremely influential hierarchies never quite agree with each other. Tantalizingly, they seem to talk about the same thing (a graded series of realities), yet their major terms never really match up. Clearly, if we could find some way that these two hierarchies were genuinely related to each other, we would have taken an important step toward the hoped-for integration of premodern and modern.

In researching this problem, I did an extensive data search of several hundred hierarchies, taken from systems theory, ecological science, Kabalah, developmental psychology, Yogachara Buddhism, moral development, biological evolution, Vedanta Hinduism, Neo-Confucianism, cosmic and stellar evolution, Hwa Yen, the Neoplatonic corpus—an entire spectrum of premodern, modern, and postmodern nests. After I had collected several hundred hierarchies, I tried grouping them in various ways, and I eventually noticed that, without exception, they all fell into one of four major types.

These four types of hierarchies—which I call *the four quadrants*—are summarized in Figure 5-1 (which is a simple

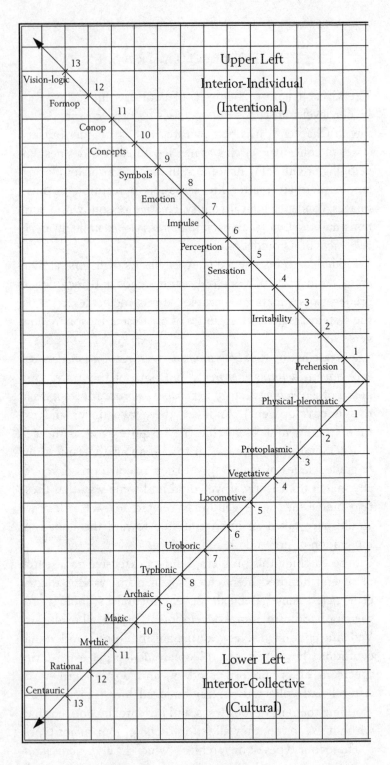

FIGURE 5-1. THE FOUR QUADRANTS

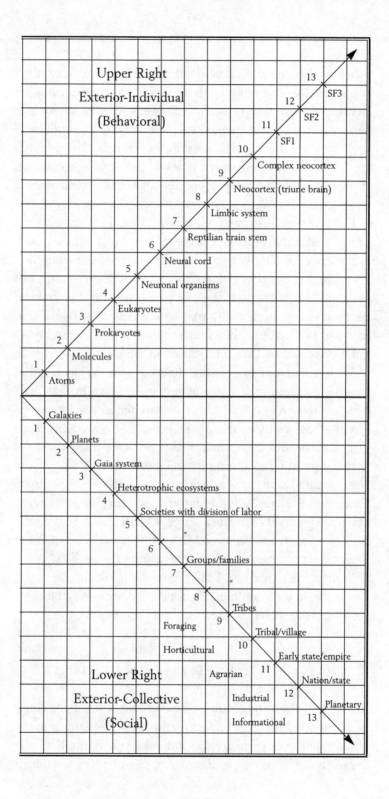

Upper Right

Exterior-Individual

(Behavioral)

13

SF3

12

SF2

11

SF1

10

Complex neocortex

9

Neocortex (triune brain)

8

Limbic system

7

Reptilian brain stem

6

Neural cord

5

Neuronal organisms

4

Eukaryotes

3

Prokaryotes

2

Molecules

1

Atoms

Galaxies

1

Planets

2

Gaia system

3

Heterotrophic ecosystems

4

Societies with division of labor

5

"

6

Groups/families

7

"

8

Tribes

9

Tribal/village

Foraging

10

Early state/empire

Horticultural

11

Nation/state

Agrarian

12

Planetary

Industrial

13

Lower Right

Exterior-Collective

(Social)

Informational

schematic, by no means complete or exhaustive, but only a representative sampling of these major hierarchies). It soon became obvious that these four different types of hierarchies simply deal with the *interior* and the *exterior* of the *individual* and the *collective* (as will soon be explained). The point is that, while these four major types of hierarchies are indeed different, they are all profoundly interrelated and deeply connected, in what look like intrinsically necessary ways.

But most fascinating of all, I found that the classic hierarchy of traditional religion and the standard hierarchy of modern science are simply two of these four types of hierarchies. As such, they are deeply interconnected with each other, but they are also part of an even larger network of hierarchical patterns. They are actually part of a universal network that involves not just two, but four, major types of hierarchies— yet hierarchies that are interrelated in vital ways.

Now, this is an interesting development. What if these quadrants, these four types of hierarchies, are in fact real? Since variations on these four hierarchies show up extensively across cultures and across epochs—premodern, modern, and postmodern—might this indicate that they are actually pointing to certain irreducible realities? What if the four quadrants are an intrinsic aspect of the Kosmos itself? Since they include *both* interior and exterior domains, might the four quadrants provide a series of crucial links in the relation of religion and science? Might they actually contain the secret key to integrating the value spheres themselves?

Perhaps, perhaps not. But it certainly looks encouraging. Let us begin by examining the four quadrants, one at a time, looking more closely at their contours.

THE OUTSIDE OF THE INDIVIDUAL

The Upper-Right quadrant is the standard scientific account of the individual components of the universe: atoms, molecules, single cells (prokaryotes and eukaryotes), multicellular organisms, including (in increasing complexity) organisms with neural cords, reptilian brain stem, paleomammalian limbic system, neocortex, and complex neocortex (with its own higher structure-functions labeled "SF1," "SF2," and "SF3").

This hierarchy shows an *asymmetrical* increase in *holistic* capacity. "Asymmetrical" means "not equivalent": atoms contain neutrons, but neutrons do not contain atoms; molecules contain atoms, but not vice versa; cells contain molecules, but not vice versa. That "not vice versa" establishes an *irreversible hierarchy of increasing wholeness*, increasing holism, increasing unity and integration. This is why all such hierarchies are indeed "higher-archies," containing successively higher or deeper or wider wholes.

Put differently, each successive unit *transcends* but *includes* its predecessors. Each senior element contains or enfolds its juniors as components in its own makeup, but then adds something *emergent*, distinctive, and defining that is not found in the lower level: it transcends and includes.

To put it one last way, each element is a *whole* that is simultaneously a *part* of another whole: a whole atom is part of a whole molecule, a whole molecule is part of a whole cell, a whole cell is part of a whole organism, and so forth. Each element is neither a whole nor a part, but a whole/part.

Arthur Koestler coined the wonderful word *holon* to refer to such "whole/parts." Virtually all natural hierarchies, in any domain, are composed of holons, wholes that are simultaneously parts of other wholes. For exactly this reason, Koestler pointed out that the word *hierarchy* should really be *holarchy*. All natural hierarchies—that is, all natural holarchies—are composed of whole/parts or holons, and they show increasing orders of wholeness, unity, and functional integration.

(Unless, of course, there is a pathology in the holarchy. Holarchies evolve and develop, as we saw, by the process of differentiation-and-integration, and if anything goes wrong with either, a pathology develops. Most antihierarchy critics confuse natural holarchy with pathological holarchy, and catastrophically end up condemning both, an error to avoid. I will use "hierarchy" and "holarchy" interchangeably when referring to their natural and normal forms, and "pathological hierarchy" or "pathological holarchy" for their aberrant displays.)

The fact that each holon is actually a whole/part places it in a profound tension: in order to exist, it must in some sense retain its own identity or its own agency as a relatively autonomous whole; yet it must *also* fit in with the other holons that are an intrinsic part of its environment. Thus, every holon must maintain not only its own *agency*, but its own *communion*, its extensive networks of relationship upon which its own existence fundamentally depends. If any holon profoundly disrupts either its agency (as a whole) or its communion (as a part), it simply ceases to exist.

THE OUTSIDE OF THE COLLECTIVE

The Upper-Right quadrant, then, is the evolutionary unfolding of individual holons according to modern science. If we now look at the *communities* or *societies* of these holons, again according to modern science, we find the Lower-Right quadrant (which I also call the *social*).

At first this quadrant might seem a bit confusing, since in the Upper-Right quadrant each higher level gets *bigger* (e.g., molecules are bigger than atoms because they contain atoms as subholons), but in the Lower-Right quadrant, each higher level gets *smaller*. This has often confused theorists who are trying to correlate various holarchies, because this holarchy seems to be running backward. What is going on here?

Erich Jantsch was one of the first to point out that, in almost any evolutionary or developmental sequence, where the *individual* holons generally get bigger (in comparison with the previous level), their collective or *communal* forms generally get smaller. The reason is twofold: One, since individual holons subsume and contain their predecessors, there will always be fewer holons the higher the level (there will *always* be fewer cells than molecules, fewer molecules than atoms, fewer atoms than quarks, etc.). And two, since there are fewer holons at each higher level, when they are gathered together into their social or collective forms, the collective will be smaller than its predecessor. Thus, as you can see in the Lower-Right quadrant of Figure 5-1, families are smaller than ecosystems, which are smaller than planets, which are smaller than galaxies.

This is generally summarized in the formula: *Evolution produces greater depth, less span.* "Depth" refers to the number of levels in the hierarchy of any holon, and "span" refers to the number of holons on that level. Each higher holon has more depth (it includes more previous holons in its own makeup), but there are fewer holons at that greater depth, and thus the collective becomes smaller and smaller—the so-called *pyramid of development.*

The Lower-Right quadrant, then, is simply a summary of the collective forms of holons as they have evolved, according to modern empirical and systems science.

THE INSIDE OF THE INDIVIDUAL

If we now look to the Upper-Left quadrant, we see yet another holarchy, this time of *interior awareness.* This holarchy moves from simple prehension to irritability (the capacity of protoplasm to respond to outside stimuli), to sensation, perception, impulse, emotion, images and symbols, concepts, concrete rules and operations ("conop"), formal-reflexive cognition ("formop"), and creative vision ("vision-logic").

This hierarchy, too, is a holarchy; it is composed of holons. Each senior holon includes, as components in its own makeup, the earlier and junior holon(s), but then adds a special and emergent capacity not found on the lower levels. Thus, each level is a whole that is part of the whole of the next higher level: each level is a whole/part, a holon, possessing both agency (wholeness) and communion (partness).

This holarchy is, of course, an *interior* holarchy, and for just that reason this entire domain was originally denied and rejected by scientific materialism, behaviorism, and positivism. The modern behavioristic claim was that mental intentionality had no reality apart from its exterior manifestation in specific observable behavior. The "mind" itself was just a "black box," unobservable by empirical science (that is, unobservable by the exterior eye of flesh) and thus not open to scientific investigation (translation: not really real). The collapse of the Kosmos included an aggressive attempt to turn all interior psychology into exterior behaviorism, and only slowly have the psychological domains fought their way back to some sort of recognition.

But for the moment, we are not trying to decide which of these quadrants is "real," or which is "important," or which is most "significant." We are simply looking at the results of a data search based on reputable investigators who have reported their findings from each of these quadrants (and thus we need to "bracket" any attempt at reductionism, if at all, until we finish the survey).

If we look at the research of the investigators of this Upper-Left quadrant, we find that what I have listed in Figure 5-1 is a fairly standard and widely accepted hierarchy, some version of which is presented by most modern developmental psychologists (from Abraham Maslow to Jean Piaget to Lawrence Kohlberg to Carol Gilligan to Jane Loevinger). Moreover, it is a hierarchy that is also quite similar, as far as it goes, to that presented by traditional and classical psychologists from Aristotle to Plotinus to Asanga to Aurobindo (as we will see in greater detail later). What they are all reporting—and generally agreeing on—are some of the basic contours of the interior of the individual, if examined closely and carefully.

Notice the difference between the interior of the individual—such as the mind—and the exterior of the individual—such as the brain. The mind is known by acquaintance; the brain, by objective description. You know your own mind directly, immediately, intimately—all the thoughts and feelings and yearnings and desires that run across your awareness moment to moment. Your brain, on the other hand, even though it is "inside" your organism, is not *interior* in your awareness, like your mind. The brain, rather, is known in an exterior and objectifying fashion; it consists of systems such as the neocortex and neurotransmitters such as dopamine, acetyl-

choline, and serotonin. But you never directly experience
something you identify as dopamine. You do not get up in
the morning and exclaim, "Wow, what a dopamine day!" In
fact, you cannot even see your brain unless you cut open
your skull and get a mirror. But you can see your mind right
now.

At the very least, then, the mind and the brain are two different views of your individual awareness, one from within,
one from without; one interior, one exterior. Each has a very
different phenomenology—they "look" quite different. The
brain looks like a crumpled pink grapefruit; the mind looks
like . . . all the joys and desires and sorrows and hopes and
fears and goals and ideas that fill your awareness from
within. No doubt the brain and mind are intimately connected—they are the Right- and Left-Hand aspects of your
individual awareness—but they also possess some profound
differences that prevent either from being reduced, without
remainder, to the other.

For the moment, then, we simply note that those researchers who have investigated the interior aspects of individual holons on their own terms are in general and broad
agreement as to the holarchy shown in the Upper-Left quadrant of Figure 5-1.

THE INSIDE OF THE COLLECTIVE

The individual holons of the Upper-Left quadrant exist in
communities, as do all holons. When individual and subjective cognitions are shared or exchanged with other individuals, the result is a collective *worldview* or communally shared
outlook. As individual cognitive holons develop and evolve—
as the awareness in individuals increases in depth from simple sensation to images to concepts to reason (the Upper
Left)—so does the collective worldview become deeper and
more complex (the Lower Left).

These *collective worldviews* are summarized in the Lower-
Left quadrant of Figure 5-1. The meaning of the terms (e.g.,
"uroboric" means reptilian, "typhonic" means paleomam-

malian) will become more apparent as we proceed. Where the Upper-Left quadrant represents individual, interior, *subjective* awareness, the Lower Left represents the collective or *intersubjective* forms of awareness, the shared *cultural* meanings, values, and contexts without which individual awareness does not develop or function at all.

This quadrant, too, represents a general consensus of serious scholars in the field who have investigated the evolution of cultural holons on their own terms. In the human realm, for example, the evolution from archaic to magic to mythic to mental has been extensively documented by scholars from the remarkable Jean Gebser to Gerald Heard to Erich Neumann to Robert Bellah to Jürgen Habermas (whom many scholars, myself included, consider the world's greatest living social philosopher).

Note that we will be referring to the inside of the collective as *cultural*, and the outside of the collective as *social*. Both are intrinsic aspects of who and what you are, but one is known from within, the other from without.

THE FOUR FACES OF THE KOSMOS

If we now look at these four quadrants and attempt to ascertain exactly how and why they fit together—what *are* these quadrants and what do they actually mean?—we soon notice that both Right-Hand quadrants represent *objective* or *exterior* realities, and both Left-Hand quadrants represent *subjective* or *interior* realities. In other words, the Right-Hand quadrants are what holons look like *from the outside*, in an objectifying, empirical, scientific type of investigation. The Left-Hand quadrants are what holons look like *from the inside*, from the interior, as part of directly lived awareness and experience.

Likewise, everything on the Right Hand has *simple location*, or location in the sensorimotor and empirical world; but nothing on the Left Hand has simple location at all, because these holons are located not in physical space but in emotional and mental and cognitive spaces (spaces of intention, not simply spaces of extension). Thus, you can point to a rock, a planet, a town, a family, an ecosystem—they all have simple location; but you cannot point to love, envy, pride, joy, or compassion: the former are exterior or Right-Hand realities, the latter are interior or Left-Hand realities.

Where the Right Hand is exterior and the Left Hand is interior, the upper half is individual and the lower half is collective or communal. Putting these all together, the four quadrants represent the exterior and the interior of the individual and the collective. In short: the intentional, behavioral, cultural, and social aspects of holons in general.

Each of these aspects, as you can see on Figure 5-1, has *correlates* with all the others. Each is intimately related to the others, for the simple reason that you cannot have an inside without an outside, or a plural without a singular. The four quadrants, I suggest, might therefore be *intrinsic aspects or features of the Kosmos itself.* Erase any one of the quadrants, and the others disappear, because they are so many sides of any given phenomenon. Exactly what all this means will, I trust, become clearer as we proceed.

As indicated, the interesting thing about all four quadrants is that they are largely uncontested by scholars working in the various fields. The sequence of atoms to molecules to cells to organisms is widely acknowledged by natural scientists. The sequence of sensation, perception, impulse, symbols, and concepts is largely agreed upon by developmental psychologists, ancient and modern. The existence of the exteriors of the collective—whether galaxies and planets or the material forms of techno-economic production (foraging to horticultural to agrarian)—is largely uncontested by serious scholars in the field. And the various worldviews (such as archaic to magic to mythic to mental) have been investigated by several renowned scholars who, despite genuine differences, present a generally similar tale of their sequence in humankind's history.

The problem, we will see, is that many scholars, specializing in only one quadrant, deny importance or even existence to the others. And this, we will see, is a direct result of the collapse of the Kosmos—of the disaster of modernity that denied reality to any of the interior dimensions at all. But if we look at the four quadrants without trying to reduce any to the others, a surprise indeed awaits us.

THE BIG THREE: I, WE, AND IT

We saw that the core of modernity was the widespread differentiation of art, morals, and science (or I, WE, and IT). But if we now look at the four quadrants, we find that they correlate exactly with these domains. The Upper-Left quadrant is described in I-language, the Lower-Left quadrant is described in we-language, and both of the Right-Hand quadrants, because they are objective exteriors, are described in it-language.

And so, in something of a surprise turn, we have arrived back at the "Big Three" cultural values spheres of art, morals, and science; the Beautiful, the Good, and the True; I, WE, and IT. Here are a few aspects of these crucial dimensions:

I (Upper Left): Consciousness, subjectivity, self, and self-expression (including art and aesthetics); truthfulness, sincerity; irreducible and immediate lived awareness; first-person accounts.

We (Lower Left): Ethics and morals, worldviews, common context, culture; intersubjective meaning, mutual understanding, appropriateness, justness; second-person accounts.

It (Right Hand): Science and technology, objective nature, empirical forms (including brain and social systems); propositional truth (singular and functional fit); objective exteriors of both individuals and systems; third-person accounts.

I refer to these three value spheres as the "Big Three" because they are three of the most significant of modernity's differentiations, destined to play a crucial role in so many areas of life. This is not simply my own idea. The Big Three are recognized by an influential number of scholars. They are Sir Karl Popper's three worlds: subjective (I), cultural (WE), and objective (IT). They are Habermas's three validity claims: subjective sincerity (I), intersubjective justness (WE), and objective truth (IT). They are Plato's Beautiful, Good, and True. They even show up in Buddhism as Buddha, Dharma, and Sangha (the I, the It, and the We of the Real, as will soon become obvious).

And of enormous historical importance, the Big Three showed up in Kant's immensely influential trilogy: *Critique of Pure Reason* (objective science), *Critique of Practical Reason* (morals), and *Critique of Judgment* (aesthetic judgment and art). Dozens of examples could be given, but that is the general picture of the Big Three, which are just a shorthand version of the four quadrants.

The fact that the four quadrants (or simply the Big Three) are the results of an extensive data search across hundreds of holarchies; the fact that they show up cross-culturally and nearly universally; the fact that they recur in philosophers from Plato to Popper; the fact that they strenuously resist being reduced or erased from consideration—ought to tell us something, ought to tell us that they are etched deeply into the being of the Kosmos, that they are the warp and woof of the fabric of the Real, announcing abiding truths about our world, about its insides and outsides, about its individual and communal forms. Ought to tell us, that is, that we are simply looking at the four faces of the Kosmos, the four corners of the known world, and none of them apparently will go away, no matter how tightly we close our eyes.

MODERNITY AND FLATLAND

We now arrive at an absolutely crucial turning point, namely, the point where the *differentiation* of the Big Three (the dignity of modernity) degenerated into the *dissociation* of the Big Three (the disaster of modernity). This dissociation allowed an explosive empirical science, coupled with rampant modes of industrial production—*both of which emphasized solely it-knowledge and it-technology*—to dominate and colonialize the other value spheres, effectively destroying them in their own terms.

Thus, the Left-Hand or *interior* dimensions were reduced

to their Right-Hand or *exterior* correlates, which utterly collapsed the Great Chain of Being, and with it, the core claims of the great wisdom traditions.

Left collapsed to Right. There, in four words, is the precise disaster of modernity, the disaster that was the "disenchantment of the world" (Weber), the "colonialization of the value spheres by science" (Habermas), the "dawn of the wasteland" (T. S. Eliot), the birth of "one-dimensional man" (Marcuse), the "desacralization of the world" (Schuon), the "disqualified universe" (Mumford).

By any other name, the disaster known as flatland.

PART II

PREVIOUS ATTEMPTS

AT INTEGRATION

6

THE REENCHANTMENT
OF THE WORLD

A map of the universe drawn by the late eighteenth and nineteenth centuries, and continuing down to today in the official mood of empirical and systems science, would essentially be nothing but the Right half of Figure 5-1. Interior holons, such as images, symbols, and concepts, were allowed no substantial reality *on their own;* they were merely *representations* of something in the Right-Hand world, the material world, which now alone was real.

Thus, in the empiricist (and behaviorist) psychology that would seize and freeze the Western soul for almost three centuries (and sophisticated versions of which are still dominant in cognitive science), the mind itself was a *tabula rasa*— a blank slate—filled with nothing but *pictures* of the sensorimotor, empirical, or Right-Hand world. There was nothing in the mind that was not first in the senses, and thus all higher modes of knowing (from the eye of mind to the eye of contemplation) were relentlessly and unsparingly reduced to empirical sensations, which is to say, they were completely destroyed in their own terms.

And so it came about, in this fractured fairy tale, that the interior dimensions of the Kosmos were simply gutted and laid out to dry in the blazing sun of the monological gaze. It is important to realize that *this was not simply or even especially an attack on spiritual realities;* it was an attack on the entire sweep of interior, introspective, lived awareness and consciousness—an attack on the Left-Hand dimensions in toto, whether "low" or "high" didn't matter in the least. *None* of those interior dimensions has simple location in the sensorimotor world, and thus *none* of them was primarily or irreducibly real.

What was real was the world of matter and energy, the world of scientific materialism. The fact that this material reality was usually held to be organized into *holistic systems of dynamically interwoven processes* did not in the least alter the fact that the systems themselves were essentially empirical, objective, positivistic, monological: in short, a flatland holism of interwoven ITs.

This meant, of course, that the entire Great Chain of Being was collapsed to its lowest level, that of empirical or sensorimotor events. For the Great Chain was, above all, the great holarchy of interior consciousness as it developed from matter to sensation to perception to images and symbols and concepts, to rational and higher rational capacities, into transrational modes of soul and spirit. (Note: Figure 5-1 represents only overall evolution up to the present, and therefore the higher modes of soul and spirit and not listed in that figure. But the entire point of "the perennial philosophy" or the great wisdom traditions—from Plato to Asanga to Plotinus to Padmasambhava and Lady Tsogyal—is that there are *higher modes of development* beyond rationality [and the eye of mind] that are disclosed in contemplation [and the eye of spirit]. These higher modes in the Upper-Left quadrant were disqualified, not because they were especially singled out, but because the entire Left-Hand dimensions of the Kosmos were equally and thoroughly rejected. We will pursue this important theme of higher development in subsequent chapters.) The point for now is that when the interior dimensions were rejected in toto, the entire Great Chain simply collapsed.

Thus the modern West became the first and only major civilization in the history of humankind to be without the Great Nest of Being. In not much more than a single century, the richly textured and multidimensional Kosmos underwent a shuddering collapse into a flat and faded system of monotonous ITs, utterly devoid of consciousness, care, compassion, concern, values, depth and Divinity.

THE DISQUALIFIED UNIVERSE

This quivering collapse of the Kosmos into nothing but Right-Hand objects and ITs was not, as earlier noted, the result of a Newtonian worldview as opposed to an Einsteinian worldview. In fact, the sciences of both Newton and Einstein (and Bohr and Planck and Heisenberg) contributed equally to this collapse by furthering the cause of monological science at the expense of the subjective and intersubjective domains. The greater the authority of physics and the natural sciences, the less real and less significant appeared the entire sweep of interior apprehensions—moral wisdom, contemplative insights, interpretive knowledge, introspective perceptions, aesthetic-expressive realities—upon which the entire Left-Hand dimensions of the Kosmos rested. The more the world stood in awe of Newton, Einstein, Kelvin, Clausius, Maxwell, Bohr, Planck, and company, the more it looked solely, even desperately, to these men and their monological know-how to deliver real knowledge and hoped-for salvation.

The sweeping success of scientific empiricism—it has dramatically dominated the worldview of modernity (so much so that even the numerous countercultural or countermodernity movements all defined themselves *in reaction to* scientific materialism)—was not due to any sort of evil intent on the part of the natural scientists themselves. By and large they were (and are) decent men and women carefully and laboriously investigating the Right-Hand dimensions of the Kosmos. But they were so stunningly successful that the other approaches—from art to morals to hermeneutics to contemplation—seemed pale and anemic by comparison. It was an embarrassment of riches, a cornucopia of truth that began through sheer exuberance to crowd out the other, softer voices in the universe.

Make no mistake, these other voices themselves contributed to this hegemony of the hardheaded with their own barely concealed envy and jealousy. That philosophy could produce real knowledge like Newton! That theology could prove Spirit with scientific precision! That God would answer the call of the laboratory! That the Goddess could be seen with a telescope! Kant was merely one of the first in an endless line of theorists to twist philosophy, psychology, and theology into a series of pretzels in order to accommodate the blinding light of Newton and Einstein and Planck and company.

And how understandable this rush to the Right Hand was! After all, every holon in the Kosmos has at least these four aspects or dimensions—behavioral, intentional, cultural, and social. Thus, *every Left-Hand event does indeed have correlates in the Right Hand.* You can see this in Figure 5-1. Wherever we find emotions, we find a limbic system. Wherever we find intentional rationality, we find a neocortex, and so on.

Thus, instead of trying to gain introspective knowledge—which, after all, is a delicate and tricky affair, and often quite hard to pin down with much certainty—let us simply investigate the brain and its empirical processes. Instead of joy, let us examine levels of dopamine. Instead of depression, let us look to serotonin at the synapses. Instead of interior angst, let us look to empirical amounts of acetylcholine in the hippocalamus. These, after all, can be *empirically seen* and *measured* with the eye of flesh. They have simple location and extension. Their results can be repeated in similar experiments. Let us therefore have done with that "introspective" nonsense and turn the entire affair of consciousness over to those variables that can be empirically and scientifically registered. Let us look to the Right-Hand world!

Still, it was not the investigation of the Right-Hand aspects of the Kosmos that caused the modern collapse. Much more than that, it was the growing belief that the entire Left-Hand dimensions were really just poorly understood Right-Hand events. A religious experience was not actually the disclosure of spiritual realities, it was simply a massive discharge of dopamine in the brain. God (or the sacred) is not needed or allowed to enter the picture at all. Likewise with compassion and love and awareness and intentionality in general: they are all *really* just Right-Hand events in the biophysical brain. In the stunning move that defined the disaster of modernity, interior states were stripped of their actual contents, because the only "real" referents (or existing entities) were those with simple location, those with Right-Hand credentials, those with an empirical passport, those mindless monological ITs. The *referents* of mental and spiritual propositions were not actual interior realities (perceived by the eye of mind or the eye of contemplation), but merely permutations on sensorimotor ITs perceived by the eye of flesh. (Those empirical correlates are very real and very important; the modern nightmare was the growing belief that those simple sensory correlates were themselves the sum total of reality.)

Thus, all of the Left-Hand and interior domains—including mind, soul, and spirit—were beginning to look more and more like hangovers from the premodern, prescientific ignorance of humankind; and a diligent, thorough, persistent examination of the empirical and positivistic realities of this world, the world of objects and ITs, would yield all the knowledge fit to know and all the salvation reality could offer.

The moment of truth of the scientific approach—a truth utterly lacking in premodern worldviews and among the Great Chain theorists—was that every Left-Hand event does indeed have a Right-Hand correlate. Transcendental events in consciousness do indeed have specific empirical correlates in the brain, a fact noted in none of the world's great religious literature. The mind itself, far from being nothing but an otherworldly soul trapped in a material body, is intimately interwoven with the biomaterial brain (not reducible to it, but not drastically divorced from it either).

Science was bound to find this out sooner or later, and this shocking discovery—Left-Hand consciousness has a Right-Hand correlate—shook to its very foundations the entire "metaphysical" approach to reality that had dominated every premodern worldview without exception. What had been thought for millennia to be radically transcendent and otherworldly was turning out to be much more immanent, this-worldly, empirical, and organic. This monological enterprise therefore made a great deal of sense—initially—because it was indeed disclosing some profound truths about what had previously been mistaken as merely "otherworldly" and "disembodied" and "metaphysical" events.

But as a confident modernity began to erase in earnest the entire Left-Hand dimensions (including the Great Holarchy), it failed to notice that this scientific endeavor was likewise erasing all sense and significance from the Kosmos itself. For there are no values, no intentions, no depths, and no meaning in any of the Right-Hand domains. The Left Hand is the home of *quality*, the Right Hand of *quantity*. The Left is the home of *intention* and thus *meaning*; the Right, of *extension* without purpose or plan. The Left has *levels of significance*; the Right has *levels of magnitude*. The Left has *better* and *worse*, the Right merely *bigger* and *smaller*.

For example, compassion is *better* than murder, but a planet is not better than a galaxy. Health is better than illness, but a mountain is not better than a river. Mutual respect is better than contempt, but an atom is not better than a photon. And thus, as you collapse the Left to the Right—as you collapse compassion to serotonin, joy to dopamine, cultural values to modes of techno-economic production, moral wisdom to technical steering problems, or contemplation to brain waves—you likewise collapse quality to quantity, value to veneer, interior to exterior, depth to surface, dignity to disaster.

The result is what Weber famously called "the disenchanted world" and Mumford so memorably called *the disqualified universe*: a world with no quality or meaning at all, ruled not by spirit or consciousness or purpose or meaning, but merely and always by blind chance or systems necessity, with the blind leading the blind.

THE POSTMODERN REVOLT
AGAINST FLATLAND

In the wake of this *modern* collapse into positivism, empiricism, behaviorism, and systems theory—all monological itendeavors—there would soon arise a series of *postmodern rebellions*, all fueled, in whole or part, by a resurgence of the interior domains screaming to be heard, acknowledged, realized, honored.

The names of these postmodern rebellions are legion (using "postmodern" in the general sense, meaning any movements occurring in the wake of modernity). Although this is a complex and intricate topic, perhaps we can say that, in a very general sense, they fall into four broad camps: Romantic, Idealist, Postmodern, and Integral.

In this and the next three chapters, we will briefly explore these reactions to flatland, all of which were also *attempts to integrate the Big Three*—which by now had disastrously *dissociated*—and thus "reenchant the world" by bringing science, spirituality, art, and morals into some sort of mutual accord.

These various approaches are not simply historical curiosities. All of them are still with us today, forming the backbone of virtually every attempt to integrate science and religion, from epistemological pluralism to ecophilosophy to postmodern paradigm. Their many successes—and their many failures—are crucial guideposts on our quest for integration.

IMMANUEL KANT AND
THE BIG THREE

Immanuel Kant was perhaps the first great philosopher to fight the leveling and deadening of the modern monological collapse, yet the net effect of his work—certainly in the hands of less gifted theorists—was to cement the positivistic hegemony, which, most scholars agree, would have been the very last thing he wanted.

Kant began by convincingly demonstrating that theoretical reason (pure reason, monological rationality, objective it-knowledge) was confined to the categories that organize sense experience. Monological it-rationality, in other words, was limited to categories of the sensorimotor domain (the Right-Hand dimensions in general), and thus pure reason was incapable of grasping, let alone proving, metaphysical or transcendental realities (such as God, freedom, and the timelessness of the soul).

Yet here were the philosophers and theologians all talking about proofs of Spirit's existence, freedom of the will, or immortality of the soul, yet none of these propositions actually had any genuine cognitive validity at all. They were all attempts by reason to step outside the realm in which it is competent, and the result was not actual knowledge, but utter and unprovable nonsense. We might say that scientific reason (it-rationality) cannot grasp God because God is not an empirical object.

Critique of Pure Reason (written in 1781) relentlessly exposed the inadequacies of monological reason to grasp metaphysical truths, and it basically marked the dramatic and historical end of that type of metaphysics. *The death of traditional metaphysics:* this was the virtually unarguable conclusion of Kant's first critique.

But for Kant, this was just the opening act. He demonstrated that monological reason cannot prove the existence of Spirit, freedom, or immortality. *But he also demonstrated that reason could not disprove their existence either.* So science was not allowed to do two things: (1) it could not say that Spirit existed; but (2) it most certainly could not say that Spirit did not exist! Kant's point was that, as he put it, he wanted to demolish knowledge (it-knowledge) in order to make room for faith. Only as objectivistic, positivistic, monological reason stopped trying to get its hands on Spirit, could other types of knowing step in to take up the fight.

Thus, in his second critique (*Critique of Practical Reason*, 1788), Kant attempted to show that where *monological reason* fails to prove (or disprove) Spirit, *dialogical reason* can succeed, at least in certain suggestive ways. For if scientific reason (it-rationality) cannot grasp God, dialogical reason (moral, ethical, practical reason) does tend to show us a type of transcendental and spiritual knowledge. Moral reason (not it-knowledge but we-knowledge) can, he believed, operate only under the assumption that Spirit exists, that freedom makes sense, and that there is a type of immortality to the soul. His argument, basically, is that the interior "ought" of moral reasoning could never get going in the first place without the postulates of a transcendental Spirit: the stomach would not hunger if food did not exist. And where monological it-knowledge can tell us precisely nothing about this spiritual domain, dialogical we-knowledge operates with its postulates all the time!

We can already see that Kant has begun to differentiate clearly the Big Three value spheres (art, morals, and science; I, WE, and IT), and he has dramatically taken spiritual knowledge out of the merely it-domain of science and placed it squarely in the we-domain of moral reasoning and yearning. He wants to limit it-science (and "it-metaphysics"), but only to make room for "we-metaphysics" and dialogical reason and spiritual faith. Morals, not science, point most clearly to God.

What remained to be done was to find some way to integrate this moral we-wisdom with scientific it-knowledge, and in his third great critique (*Critique of Judgment*, 1790), Kant attempts this integration, in part through the expressive-aesthetic dimension (or art in the most general sense). In other words, he wants to introduce the aesthetic I-domain in order to integrate we-morals and it-science. *He wants to integrate the Big Three.*

THE WESTERN WATERSHED

Here we reach an absolutely crucial turning point for the Western world, the very divide between the modern and the postmodern moods. These types of categorizations are always slippery, but it might fairly be said that Kant was either the last of the great modern philosophers or the first of the great postmodern philosophers. He probably was both. But in any event his work is a branch from which stem, in whole or part, virtually all of the four camps: Romantic, Idealistic, Postmodern Poststructuralist, and Integral.

You can easily see that, depending on which of Kant's three critiques you emphasize, you can extract a dramatically different worldview from this great man's work. If you focus on *Critique of Pure Reason*, you could readily become a dedicated positivist and behaviorist: science alone gives cognitive knowledge, "real" knowledge, and all else is nonsensical metaphysics. Let us therefore confine ourselves to the study of sensorimotor phenomena (like Newton!), and relegate everything else to the dustbin of meaningless metaphysics. And indeed, many of the positivistic and antimetaphysical currents in the West trace their lineage directly to Kant's first critique.

But if you focus on the second critique (*Practical Reason*), you will have a very different story to tell. Science delivers genuine it-knowledge, but who cares? The real action is in the moral yearning and ethical reasoning that, if they do not disclose, nonetheless powerfully indicate, spiritual realities. Men and women are not free as *empirical objects*—in the world of ITs, there are only causality and determination (whether strict or statistical). But as *ethical subjects*, men and women are indeed autonomous, or can be if they rise to their own highest occasion and act according to a universal, worldcentric, moral reasoning: not what is right for me and my tribe, or me and my mythic religion, or me and my nation, but what is right and fair for all peoples regardless of race or creed. For when I act in this worldcentric—not egocentric, not ethnocentric, but worldcentric—fashion, I am free in the deepest sense, for I am obeying not an outside force but the interior force of my own ethical reasoning: I am autonomous, I am deeply free.

And that was the exhilarating message of Kant's second critique. It doesn't matter if the world of ITs is a deterministic system, because in the moral stance of worldcentric ethical embrace, I am a free soul, free because those dictates issue from my own deepest being. Numerous religious, spiritual, and especially ethical theories would trace their lineage to this extraordinary second critique. In fact, to this day, many of the great moral theorists from Rawls to Habermas would be described as "Neo-Kantian."

If you focused on the third critique, yet another stunning story would emerge. Granted, science yields genuine knowledge of ITs; and granted, we-morals open us to a spiritual wisdom. But how do we integrate these separate realms? And wouldn't that integration actually be *the* highest and most desirable goal? And if ART is the great bridge between science and morals, is not world salvation in the hands of the artists?

Well, many artists thought so; and in the wake of Kant's third critique (not to mention the French Revolution), the great Romantic *aesthetic-expressive* movements of modernity and postmodernity began, movements that would locate ultimate reality not in the it-domain of science or the we-domain of morals, but in the I-domain, the subjective domain, the domain of art and artistic vision and intense self-expression. Not just Truth, not just Goodness, but above all Beauty, would finally disclose the Divine. And these great aesthetic-expressive movements began in earnest with the Romantics of the late eighteenth century.

The extraordinary attempt to reenchant the world had just begun.

ROMANTICISM:

RETURN OF THE ORIGIN

Kant's final goal—to integrate the Big Three of art, morals, and science—ultimately eluded him. Despite his heroic attempts in the third critique to achieve this integration via art and organic telos, most theorists agree that he failed. It is certain that the theorists in his immediate wake believed that he failed, for they took up the task with an astonishing vigor. The simplest way to state this failure is that art could not itself achieve the integration because it was merely one of the three spheres to be integrated, and thus it could not itself accomplish the job.

But that was a fact that the Romantics in general failed to recognize or chose to ignore, and they began an intense effort to make the I-domain, the subjective domain—and especially the domain of aesthetics, sentiment, emotion, heroic self-expression, and feeling—the royal road to Spirit and the Absolute.

THE PRE/TRANS FALLACY

Kant had spotted, and indeed was part of, the extraordinary dignity of modernity, in that he had clearly differentiated the Big Three value spheres of art, morals, and science. But he also realized that the Big Three were starting to fly apart—not just differentiate but dissociate—and that monological it-science was taking advantage of this fragmentation to begin its imperialistic adventures. Kant is *already* trying to beat back it-science "in order to make room for faith." And he is *already* trying to pull the Big Three together in his third critique. But try as he might, he cannot effect the sought-after integration. The Big Three are dissociating; Kant knows it; and he is powerless to prevent the fragmentation or the "diremption," as they were already calling it.

The Romantics took their own approach to this fragmentation and dissociation, but an approach that, it turned out, was in some ways significantly flawed despite the best intentions. As we saw earlier, if you confuse differentiation and dissociation—it's an easy mistake—then you will attempt to cure the dissociation by getting rid of the differentiation itself. You will push back in time, not prior to the dissociation—which is correct—but prior to the differentiation—which is simply wholesale regression. You will try to push back to some sort of prior *fusion* or *undifferentiated* state, some sort of "primal" and "pristine" and "pure" state, prior to the madness of modernity altogether. You will want to get back to nature, back to the noble savage, back to the purity and innocence of a primal past. You will be a retro-Romantic, longing for the "wholeness" and "union" of yesteryear, and ignoring any unpleasantness you might actually find in the halls of premodernity.

Thus, even today, a well-respected reference book such as *The New Columbia Encyclopedia* summarizes the general Romantic movement thus: "The basic aims of romanticism were various: a return to nature and to belief in the goodness of man, most notably expressed by Jean Jacques Rousseau— with the subsequent cult of 'the noble savage,' attention to the 'simple peasant,' and admiration of the violently self-centered 'hero'; the rediscovery of the artist [and aesthetic-expressive self] as a supremely individual creator; the exaltation of the senses and emotions over reason and intellect. In addition, romanticism was a philosophical revolt against rationalism."

Now, if you are in a revolt *against* rationality, it is rather hard to sincerely *integrate* rationality—and thus a genuine integration of the Big Three value spheres will tend to elude you. In fact, the Romantics fell violent prey to what I have called *the pre/trans fallacy,* namely, the confusion of prerational with transrational simply because both are nonrational.

Granted, spirituality is, in some sense, beyond mere rationality. But there is *trans*-rational, and there is *pre*-rational. Prerationality includes all of the modes leading up to rationality (such as sensation, vital life feeling, bodily emotion, and organic sentiment), and, by its very nature, tends to exclude rationality, no matter what lip service it might give to it. Transrationality, on the other hand, lies on the other side of reason. Once reason has emerged and consolidated, consciousness can continue to grow and develop and evolve, moving into transrational, transpersonal, and supraindividual modes of awareness. Transrationality, unlike prerationality, happily incorporates the rational perspective, and then adds its own defining characteristics; it is thus *never* antireason, but, in a friendly way, transreason.

Assuming, for the moment, that these higher dimensions exist, we can see that the overall arc of consciousness evolution and development moves from prerational to rational to transrational; from subconscious to self-conscious to superconscious; from prepersonal to personal to transpersonal; from id to ego to God.

The pre/trans fallacy occurs when the pre and trans states are confused or equated, and it operates in both directions. For example, Freud tended to take all genuine transrational experiences and reduce them to prerational infantilisms (to primary narcissism, oceanic indissociation, preambivalent oral stage, and so on). Jung, on the other hand, often erred in the opposite direction, taking some very prerational childhood productions and elevating them to transrational glory. Both of these mistakes—*reductionism* and *elevationism*—rest on a prior confusion of pre and trans.

And the Romantics were about to run into the elevationist error with a vengeance, eulogizing the prerational domains with such intensity that they often ended up in blatantly regressive nightmares. Yet it all started so nobly, so understandably, so sincerely. . . .

TO REWEAVE THE WEB OF LIFE

Prerationality, as we said, includes all of the modes of awareness leading up to formal rationality, such as sensation, emotion, imagery, and intense feeling (all of these are shown in
Figure 5-1). As rationality itself then emerges and develops, it
ideally transcends and includes, goes beyond but incorporates, the prerational domains (since, as we have seen, "transcend and include" or "differentiate and integrate" is the core
dynamic of *all* stages of normal development and evolution).

But if there is a *pathology*—if reason does not just *differentiate*, but instead *dissociates*, from the lower realms—the result
is repression and alienation, the suffocation of vital life, feeling, and emotion. Instead of transcend and include, there is
deny and repress.

If this pathological dissociation occurs, then reason, with
all its rich capacities for dialogue, ethics, mutual recognition,
and care, becomes dry and abstract and life-denying in the
worst sense. This repression is not something *inherent* in reason and reasonableness; it is a pathological aberration of reason, occurring when its necessary differentiations go too far
into morbid dissociations.

But that is exactly what was happening to modernity in
general, was it not? The rationality of modernity had ad-
mirably differentiated the Big Three of self (I), culture (WE),
and nature (IT); but now modernity, hypnotized by a suggestive scientism, was not *integrating* those realms, it was in the
process of *dissociating* those realms, with self and culture and
nature all at one another's throats, and monological it-science
colonializing the entire lot.

And one of the oppressed realms was that of aesthetics,
self, and self-expression, including all of the rich feelings,
emotions, and vital life that, being part of the Left-Hand or
interior domains, had been marginalized from serious discourse—by any other name oppressed, denied, denigrated,
devalued. In short: reason was repressing feeling.

(It is no accident that at precisely this point, the likes of
Schopenhauer, Nietzsche, and Freud would come forward to
point out this epidemic mental repression of instinctual life.

It wasn't that this type of repression had not occurred in pre-modern cultures, for almost any higher level can repress any lower level at any given time; but never had such a powerful rationality so violently clamped down on interior life, which was the essence of a dissociated modernity, itself Dr. Freud's real patient.)

The Romantics were understandably, and rightly, horrified by this repression and dissociation. And the various Romantics—Rousseau, Herder, the Schlegels, Schiller, Novalis, Coleridge, Keats, Wordsworth, Whitman—took it upon themselves to heal this violent fragmentation, not with abstract rationality, but with intense feeling—what Wordsworth called "the spontaneous overflow of powerful feelings." Herder was explicit: "See the whole of nature, behold the great analogy of creation. Everything feels itself and its like, life reverberates with life. . . . Impulse is the driving force of our existence, and it must remain this even in our noblest knowings. Love is the noblest form of knowing, as it is the noblest feeling." As for those who believed that abstract it-rationality was the only true form of knowledge, they must be, said Herder, either "liars or enervated beings."

Moreover, as one scholar of the period summarized the central Romantic aspiration for a *unified feeling of life*, "This feeling cannot stop at the boundary of my self; it has to be open to the great current of life that flows across it. It is this greater current, and not just the current of my own body, which has to be united with higher aspiration . . . if there is to be unity in the self. Thus our self-feeling must be continuous with our feeling for this larger current of life which flows through us and of which we are a part; this current must nourish us not only physically but spiritually as well." That is not some present-day New Age nostrum; it was the precise credo of the general Romantic movement, which began almost two hundred years ago (the New Age movement itself being merely one of its many descendants).

We can see that the Romantics were already trying to integrate the Big Three of self, culture, and nature, to unify that which the disaster of modernity had put asunder. For above all else, the Romantics yearned for *unity* and *wholeness*. As

Charles Taylor points out, "There was a passionate demand for unity and wholeness. The [Romantics] bitterly reproached the Enlightenment thinkers for having dissected man and hence distorted the true image of human life in objectifying human nature [reducing it to Right-Hand objects]. All these dichotomies [and dissociations] distorted the true nature of man which had rather to be seen as a single stream of life, or on the model of a work of art [the aesthetic-expressive dimension], in which no part could be defined in abstraction from the others. These distinctions thus were seen as abstractions from reality. But they were more than that, they were mutilations of man. . . . It was a denial of the life of the subject, his communion with nature and his self-expression in his own natural being."

Back to nature, back to some sort of union or communion prior to the modern fragmentation and collapse. As one historian put it, "What they [the Romantics] themselves yearned for was unity with self and communion with nature—that man be *united in communion with nature.*" This was to be accomplished by a "sympathetic insertion into the great stream of life of which we are a part"—a oneness with the great Web of Life.

This extraordinary attempt to integrate the Big Three of self, culture, and nature, and thus introduce some measure of wholeness and unity into a modernity fast becoming sick with its own conceits, was an aspiration as noble as any that can be conceived. This is why, I believe, to this day, we owe the Romantics an undying debt of gratitude. They were the first to spot the disease, more than two hundred years ago. They were the first to react to it with authentic horror. They were the first to attempt to reweave the fragments, heal the wounds, become at home in the universe, be a humble part of life's wondrous web, and not its arrogant master.

THE SLIDE

Alas, in their understandable zeal to get beyond rationality to a genuine spiritual wholeness, the Romantics often ended up recommending *anything nonrational*, including many things that were frankly *prerational*, regressive, egocentric, and narcissistic. They all too often confused prerational impulse with transrational insight; preconventional nature with postconventional spirit; preverbal expression with transverbal awareness; preconventional and egocentric license with postconventional and worldcentric freedom; and predifferentiated fusion with transdifferentiated integration.

In other words, precisely because they confused differentiation with dissociation, they confused prerational with transrational, and they set out to glorify every prerational, preconventional, preconceptual, and "natural" impulse they could find. Put bluntly, the Romantics tended to dedifferentiate, not transdifferentiate. They inadvertently eulogized fusion, not actual integration. They let self-expression slip into self-obsession and "divine egoism." And in this regressive and narcissistic slide into anything preconventional, they imperiled not just the disasters of modernity, but the dignities as well.

No wonder that so many cultural critics, from Robert Bellah to Colin Campbell to Jürgen Habermas, have seen our present-day obsession with self, sentiment, impulsive gratification, "be here now," "lose your mind and come to your senses," the white middle-class consumption of indigenous tribal religions as "pure and innocent and whole," the belief that "you create your own reality," intense sensory gratification, consumerism, self-glorification, and consequent social alienation—as being, in significant ways, direct descendants of Romanticism.

Of course, the more sophisticated Romantics never recommended pure and unadulterated regression. Rather, the idea was that we would somehow recontact and regain the "lost wholeness" but now on a "higher level" or in a "mature form," thus uniting the best of premodernity and modernity. This is assuredly a noble goal, and one that other approaches, including the Integral, would embrace.

But in practice as well as in detailed theory, the Romantics could not actually effect this integration of premodern and modern (or the integration of the Big Three). They had so devalued the rational, conventional, and bourgeois spheres that the promised "integration" of these spheres was, at best, lip service (as the despised spheres themselves were quick to point out). The fact remained that, in confusing differentiation and dissociation—and thus prerational and transrational—the Romantics often ended up with a blanket call for dedifferentiation, a process that, when it occurs in a living system, is called "cancer": a regressive dedifferentiation of cells growing out of control, ending in the death of the system.

Indeed, in this spiraling regressive yearning, you very well might become somewhat dedifferentiated yourself, finding your own ego to be the source and creator of all reality (as preoperational thinking does). Divine egoism will increasingly rear its narcissistic head, and you might be pulled, with every good intention, into the unending drama of your subjective inclinations. The world will become darker and darker, full of malevolent intent; you alone seem pure and clean in a world that does not care. You might become sadder and sadder, sick with the world's sorrow, too beautiful, really, for this wretched world. And if you are a true Romantic, you will nobly end it all with a terribly beautiful suicide. (Many of the great Romantic narratives, and many of the great Romantics themselves, ended in suicide.)

In the meantime, the search was on for the past paradise of wonderful wholeness and pristine purity that modernity had viciously destroyed. The search was for a period not just before the dissociations of modernity but before the differentiations themselves (since the two were thoroughly confused). The oak was somehow a hideous violation of the acorn; and the acorn, not the oak, possessed "more unity"—an utter confusion, to be sure, but the confusion upon which the retro-Romantics, then and now, rested their case.

Thus *the recovery of Origin* became the great theme of this period: a burning desire to find, recontact, resurrect, and embrace a lost and found Beloved, the return of the wondrous God or Goddess, which had once been gloriously present in an actual past era, but had been bruised, banished, burned, or buried by a cruel and uncaring modernity. The attempt to re-contact humanity's acornness had just begun.

THE WAY BACK MACHINE

Thus started the search for the period in history or prehistory where the terrible differentiations of modernity had not yet occurred. The Romantics had jumped wholeheartedly aboard the Regress Express, and by far the most popular destination for the early Romantics was ancient Greece.

There are, of course, numerous aspects of classical Greece that deserve much admiration, not the least of which was its precocious embrace of reason and thus its preliminary differentiation of the Good, the True, and the Beautiful (a differentiation that reason alone discloses; this differentiation—and dignity—is lacking in all prerational modes). But precisely because this differentiation was preliminary, none of the massive dissociations of modernity had yet set in; thus there tended to be a marvelous *harmony* among the value spheres in Greek thought. I think this harmony is what many people, to this day, find so attractive about classical Greece, and it certainly attracted the early Romantics.

But had the Greeks actually and fully differentiated the Big Three, they would have evidenced its fruits: they would have banished slavery (one out of every three people in this "democracy" was a slave), and they would have set into motion the apparatus for women's rights, among other dignities. To eulogize a society where many people were slaves, and women and children might as well have been, evidences, to put it mildly, a warped sense of values.

Modern-day Romantics have realized this, and almost to a one they have abandoned ancient Greece (often with horror) and pushed back even further into prehistory in search of their primal paradise—with the result, of course, that they make Greece itself the beginning of the modern disaster, and heap upon it a scorn that the early Romantics would have found incomprehensible.

For the ecofeminists, the especially hallowed period is that immediately preceding agrarian Greece, namely, the horticultural societies that flourished from roughly 10,000 B.C.E. to 4000 B.C.E., before the rise of the early empires and the agrarian "patriarchy" in general.

In *horticultural* societies, the major means of production is a simple digging stick or handheld hoe, whereas in *agrarian* societies (such as Greece), it is the heavy animal-drawn plow. Pregnant women can easily handle a hoe, whereas if they participate in heavy plowing they suffer a significantly higher rate of miscarriage. Thus, in horticultural societies, women were usually a crucial segment of the productive force. Indeed, up to 80 percent of foodstuffs in these societies were produced by females, and the social relations and mythic divinities of these cultures appropriately reflected that fact. About one third of these societies had female-only deities and about another third had male-and-female deities (and about a third, male-only deities).

It is easy to see the attraction these horticultural societies have for ecofeminists, which is perhaps why they studiously overlook the fact that 44 percent of these societies engaged in frequent warfare and more than 50 percent in intermittent warfare (so much for the peace-loving Great Mother societies); that 61 percent had private property rights (so much for communal sharing); that 14 percent had slavery (so much for slavery's being introduced by patriarchy); and that 45 percent had bride-price (so much for equal rights). These horticultural societies were anything but "pure and pristine," and if they were in touch with nature, it was a nature whose values no ecofeminist today would actually defend.

Leave it to the ecomasculinists (deep ecologists) to push back even further into prehistory, to the previous stage of *foraging*, beyond which one cannot go (prior to foraging were apes). This *must* be the pure and pristine and "nondissociated" state, because there is no further destination left on the Regress Express.

The ecofeminists have embraced horticultural matrifocal cultures as the pure and pristine state, "one with nature" in the seasonal cycles of the moon, planting, and harvesting, and consequently *condemned* the rise of patriarchal agrarian societies (e.g., classical Greece) as the fall of humanity in general. Just so, when the ecomasculinists *pushed back even further* into foraging, they *condemned* horticultural societies—the heaven of the ecofeminists—as being the first great rape of the land and the destruction of paradise. For, according to the ecomasculinists, *farming itself* is an attempt to control and dominate the purity and spontaneity of nature. Foraging, gathering, occasionally hunting what nature offered—now *that* is pure and pristine. Humankind's woes all began when a woman first took up a hoe.

And let us ignore the data that show that 10 percent of these foraging societies had slavery, 37 percent had brideprice, and 58 percent engaged in frequent or intermittent warfare. This *must* be the pure and pristine state, because there is nowhere further back to go!

Thus we can begin to see what so many of the retro-Romantic, ecoholistic, back-to-nature, recovery-of-Origin approaches have in common: what might be called the "pick-and-choose" approach to history. Pick those things you admire about a premodern epoch and studiously ignore everything else, as if pieces of the fabric of culture could be cut and pasted onto the modern world to effect the desired "integration." Compare the best of yesterday with the worst of today, and scream, "Devolution!"

Even Foucault—no great fan of modernity—was horrified at this pick-and-choose paradise: "I think that there is a widespread and facile tendency, which one should combat, to designate that which has just occurred [i.e., modernity] as the primary enemy, as if this were always the principal form of oppression from which one had to liberate oneself. Now this simple attitude entails a number of dangerous consequences: first, an inclination to seek out some cheap form of archaism or some imaginary past forms of happiness that people did not, in fact, have at all. There is in this hatred of the present [modernity] a dangerous tendency to invoke a completely mythical past."

As we will see, numerous points in the Romantic orientation are in fact quite valuable and should definitely be brought to the integrative table. We do indeed need to recontact and integrate nature, which was, horrifyingly, one of the casualties of the modern dissociation.

But premodern societies did not actually integrate self, culture, and nature; they simply had not yet fully differentiated them in the first place. They were largely *pre*-differentiated, not *trans*-differentiated, and therefore they *cannot serve as cogent models for the integration of the Big Three.* This differentiation (and its possible integration) is an *emergent*, something new and novel in the evolutionary stream. It *never* existed before (consciously or unconsciously), and therefore no amount of "return to historical Origin" will help with this novel emergent. To return to Origin forever is to miss the point forever.

Thus, just as the acorn does not actually integrate the leaves and branches and roots—for those have not yet emerged—so premodern cultures did not integrate the modern value spheres, for those had not yet fully differentiated. As with the acorn, these premodern states actually had less differentiation, less integration, less unity, less wholeness; they lacked many of the diseases of modernity because they lacked the differentiations as well. If we fail to grasp that elemental distinction, fusion is confused with integration, and the regressive slide is under way.

It is to tomorrow, not yesterday, that our vision must be turned. And Idealism began in part with exactly that realization, and exactly that attack on Romanticism. The God of tomorrow, not the God of yesterday, comes to announce our liberation.

8

IDEALISM:

THE GOD THAT IS TO COME

One of the most astonishing and radical differences between premodern and modern cultures is the *direction* in which the universe is said to be unfolding. In most premodern religions, the tale is told of the "time before time," the time of creation, where a Great Spirit of one sort or another created the world out of itself, or out of some *prima materia*, or out of nothing. In the immediate wake of this genesis, men and women, as part of that remarkable creation, lived in peace and harmony with themselves and with all other creatures. Living close to Source, close to Spirit, close to God and Goddess, humans bathed in that primordial delight and radiated goodness in all directions.

But then, it is said, through a series of strange events, either this God began slowly to withdraw from humans, or humans withdrew from this God. Either gradually or suddenly, but always and terribly, humans lost touch with the primal Eden.

In the Hindu version, the world then devolved through four yugas, or cosmic epochs, with each one becoming increasingly dark, alienated, fractured, and painful. These epochs are likened to gold, silver, bronze, and iron, leading from pure dharma (spiritual Truth) to complete adharma (spiritual wasteland); and today we are living in the corrupt iron, or Kali, yuga, farthest from the Source.

Scholars of this almost universal tale in premodern cultures summarize what it tells us about the basic form of the universe's direction: from the Age of Myth to the Age of Heroes to the Age of Men to the Age of Chaos, a steady and dismal downhill slide. Once again, we of today live in the Age of Chaos, farthest from the Source and Origin.

In all these tales, the overall direction of the universe's unfolding is unmistakable: as if following some second law of religious thermodynamics, the spiritual universe is running down. In the actual unfolding of the universe's history, we humans (and all creatures) were once close to Spirit, one with Spirit, immersed in Spirit, right here on earth. But through a series of separations, dualisms, sins, or contractions, Spirit became less and less available, less and less obvious, less and less present. *Deus abscondus:* history itself is the story of spiritual abandonment, with each era becoming darker and more sinister and less spiritual. For premodern cultures, in short, history is devolution.

But sometime in the modern era—it is almost impossible to pinpoint exactly—the idea of history as devolution (or a fall from God) was slowly replaced by the idea of history as evolution (or a growth toward God). We see it explicitly in Friedrich Schelling (1775–1854); Georg Hegel (1770–1831) propounded the doctrine with a genius rarely equaled; Herbert Spencer (1820–1903) made evolution a universal law; and his friend Charles Darwin (1809–1882) applied it to biology. We then find it appearing in Sri Aurobindo (1872–1950), who gave perhaps its most accurate and profound spiritual context, and Pierre Teilhard de Chardin (1881–1955), who made it famous in the West.

Suddenly, within the span of a mere century or so, serious minds were entertaining a notion that premodern cultures, for the most part, had never even once considered, namely that—like all other living systems—we humans are in the process of *growing toward our own highest potential*, and if that highest potential is God, then we are growing toward our own Godhood.

And, this extraordinary view continued, *evolution* in general is nothing but the growth and development toward that consummate potential, that *summum bonum*, that *ens perfectissimus*, that highest Ground and Goal of our own deepest nature. Evolution is simply Spirit-in-action, God in the making, and that making is destined to carry all of us straight to the Divine.

THE RISE OF IDEALISM

This idea—cosmic and human history is most profoundly the evolution and development of Spirit—occurred immediately in the wake of Kant, and was one of the great announcements of the Idealists. This was during that extraordinary period when the Big Three (art, morals, and science) had been clearly differentiated (around the end of the eighteenth century), but before their massive dissociation and eventual collapse (around the end of the nineteenth century). As such, this was a truly fertile period for the value spheres to enrich one another. Although the spheres had not yet been fully integrated (a task that still eludes the West), nonetheless they were all on speaking terms, perhaps the last time in Western history that such fruitful exchange occurred. Out of that astonishingly rich soil, grew Idealism.

As usual, it began with Immanuel Kant, who had famously maintained that we can never know "the thing in itself," only the appearance or phenomenon that results when the thing in itself is acted on by the categories of the human mind. German Idealism began, in a sense, with that notion, the notion that the world is not merely *perceived* but *constructed*. Not naïve empiricism, but mental idealism, has a hand in the perception of the world.

Johann Fichte, a contemporary of Kant, pointed out that if you cannot know anything at all about the thing in itself, you cannot know it exists, either. It is an utterly useless notion. At the same time, Kant had shown that phenomena are constructed by the mind. If we get rid of the impossible notion of the thing in itself, the result is that the entire perceived universe is the product of mind. Yet this obviously cannot be an *individual* mind or self—Mrs. Smith of Boise, Idaho, is obviously not producing the entire Kosmos. It must be a mind beyond you or me or any particular individual: it must be a *supraindividual* and absolute Self producing the entire universe.

This absolute Self Fichte proposed as the first principle of philosophy, and from this transpersonal Self he would attempt to derive the entire manifest universe (and in a fashion strikingly similar to the great Vedanta Hinduism of the East. For both of them, out of the absolute Self's creative imagination issues forth the finite world, and in reaction to the finite world grows the finite self. For both of them, liberation consists in rediscovering the absolute Self of which the finite self and finite world are but a manifestation).

Because all forms of knowledge (including it-science and we-morals) issued forth from this absolute Self, all forms of knowledge could, Fichte believed, be seamlessly integrated in this Self awareness, and this integration would heal the "diremption" or fragmentation of modernity, which was already beginning to rear its pathological head.

In other words, it comes as no surprise that Fichte, too, wants to integrate the Big Three. This integration, we have seen, is actually the single greatest task confronting the postmodern world. What modernity put asunder, postmodernity must heal. And the great theorists in the wake of Kant can all be situated in relation to that burning question: Now that we have successfully differentiated the Big Three, how can they be integrated? (Romanticism tried to do so by regression and dedifferentiation, a suicidal dead end. Idealism would attempt to do so by heading in the opposite direction: higher development.)

Because the absolute Self (which is Spirit itself) gives rise to the entire manifest world, Fichte maintained that part of the task of philosophy was to reconstruct what he called the "pragmatic history of consciousness"—that is, to reconstruct the actual path that consciousness has taken in its creative unfolding of the universe. Fichte was thus one of the very first to introduce the absolutely crucial and historically world-shaking notion of *development* (or evolution). The world is not static and pregiven; it develops, it evolves, it takes on different forms as Spirit unfolds the universe.

And, the Idealists maintained, understanding this unfold-

ing or development is the secret key to understanding Spirit
itself.

EVOLUTION AS SPIRIT-IN-ACTION

Friedrich Schelling took that initial developmental insight
and worked it into a profound philosophy of spiritual unfolding, and Georg Hegel hammered out its details in a series
of brilliantly difficult treatises. The general points may be
summarized as follows.

Absolute Spirit is the fundamental reality. But in order to
create the world, the Absolute manifests itself, or goes out of
itself—in a sense, the Absolute forgets itself and empties itself into creation (although never really ceasing to be itself).
Thus the world is created as a "falling away" from Spirit, as a
"self-alienation" of Spirit, although the Fall is never anything
but a play of Spirit itself.

Having "fallen" into the manifest and material world, Spirit
begins the process of returning to itself, and this process of the
return of Spirit to Spirit is simply development or evolution
itself. The original "descent" (or involution) is a forgetting, a
fall, a *self-alienation* of Spirit; and the reverse movement of
"ascent" (or evolution) is thus the self-remembering and *self-
actualization* of Spirit. And yet, the Idealists emphasized, all of
Spirit is fully present at each and every stage of evolution as
the *process* of evolution itself.

When Spirit first goes out of itself to create the manifest
universe, the result is Nature, which Schelling calls "slumbering Spirit" and Hegel calls "God in its otherness." Nature is a
direct manifestation of Spirit, and thus Nature is sacred to
the core; but it is *slumbering* Spirit, simply because Nature is
not yet self-reflexively aware. It is the lowest form of Spirit,
but a form of Spirit nonetheless. It is Spirit in its *objective*
manifestation, what Plato had called "a visible God" (or visible Goddess).

In the second major stage of development, Spirit evolves
from objective Nature to subjective Mind. Thus, Spirit has

now developed from *subconscious* to *self-conscious* (or preper-sonal to personal, or prerational to rational), and thus begins to reflect on its own existence. Where Nature was *objective Spirit*, Mind is *subjective Spirit*, and thus we see increasingly more conscious forms of Spirit's own self-actualization and return to itself.

But it is at this point that the subject and the object, or Mind and Nature, can not just differentiate but dissociate, and thus this stage is often marked by a rampant dualism—a "spiritual pathology," according to Schelling, the "unhappy consciousness," as Hegel put it. This unhappiness is not present in the previous stage of Nature, but only because Nature is slumbering; yet with the self-conscious awakening of Mind, these painful divisions become all too obvious.

Here the Idealists—especially Fichte and Hegel—veered sharply away from the Romantics, who by and large wanted to heal the painful unhappy consciousness by a "return to Nature." But this return, the Idealists pointed out, is based on a series of profound confusions. Indeed, some of the earliest, most venomous, most polemical—and altogether most accurate—critiques of retro-Romanticism came from the Idealists, who quickly crawled out of that bed and for good measure set it on fire. Fichte and Hegel rail against the Romantic regression to sentiments, feeling, antirationalism, and organic immersion, pointing out, quite correctly, that the Romantics were headed in precisely the wrong direction.

What the Idealists understood—and what I have called the "pre/trans fallacy"—is that prerational modes can appear to be transrational simply because both are nonrational. And, as we saw, in their understandable rush to go transrational, the Romantics often ended up glorifying *anything* that was nonrational, including states that were frankly regressive, narcissistic, indissociated, and dedifferentiated, all of which thoroughly erased not just the disasters of modernity but the dignities as well. This regressive catastrophe set Fichte and Hegel and occasionally Schelling on polemical fire, and rightly so. That these Idealists were witness to this regressive nightmare as it actually unfolded makes their polemics all the more cogent—and applicable to similar regressive slides now widely occurring under the guise of a "new age" and a "new paradigm."

Fichte, Schelling, and Hegel were united: there is no going back to recontact a lost Spirit—for in the "backward" direction there is only *slumbering* Spirit, which is *already* self-alienated. (This is simply another way of pointing out that the earlier stages of human development, whether phylogenetic or ontogenetic, offer no substantial models for the healing of the dissociations of modernity.)

No, it is not by a "return to Nature" that humans can end their alienation and unhappy consciousness, but rather by moving forward to the third great stage of development and evolution, that of nondual Spirit. Thus, for both Schelling and Hegel, Spirit goes out of itself to produce objective Nature, awakens to itself in subjective Mind, then recovers itself in pure nondual Spirit, where subject and object are one pure act of nondual consciousness that unifies both Nature and Mind in realized Spirit.

Thus, Spirit knows itself objectively as Nature; knows itself subjectively as Mind; and knows itself absolutely as Spirit—the Source, the Summit, the Ground, and the Process of the entire ordeal.

Note, then, the overall sequence of development: from nature to humanity to divinity; from subconscious to self-conscious to superconscious; from prepersonal to personal to transpersonal; from id to ego to God. But Spirit is nonetheless fully present at each and every stage as the *evolutionary process itself*: Spirit is the process of its own self-actualization and self-unfolding; its being is its own becoming; its Goal is the Path itself.

Thus, humans can end their alienated and unhappy consciousness, not primarily by going back to Nature but by going forward to nondual Spirit. Not preconventional Nature but postconventional Spirit holds the key to overcoming alienation and dissociation, and that Spirit is contacted, not by spiraling regression to preconventional slumber, but by evolutionary progression to a radiant Nonduality.

(Of course, when the Mind emerges, it can indeed repress

Nature, precisely because Mind is a higher-order holon that can arrogantly usurp its role in the normal holarchy by oppressing its junior holons, including Nature—a suicidal repression in that these are elements of its own being, which is why the ecological crisis is indeed suicidal. Likewise, internally, the ego can repress the id, the same dissociation as Mind repressing Nature. Under this pathological twist, the Mind must recontact Nature and befriend Nature, just as, internally, there must be "regression in service of the ego." That "befriending" is well and good, and mandatory for healing. But that is just the first step. For Mind and Nature to be genuinely integrated and unified, a third term is required, above both Nature and Mind and reducible to neither; that term, of course, is Spirit. Thus, the great integration can never be achieved by Nature alone, or by Mind alone, or by any combination of the two. Only Spirit itself, which is beyond any feelings of Nature and beyond any thoughts of Mind, can effect this radical unity. Spirit alone transcends and includes Mind and Nature. Under the pre/trans fallacy, the Romantics all too often confused preconventional Nature with postconventional Spirit, and thought that *a simple union of prerational Nature with rational Mind would be the same as transrational Spirit*, and therein was their Waterloo. For prerational Nature can be seen with the eye of flesh and rational Mind can be seen with the eye of reason, but transrational Spirit can be seen only with the eye of contemplation, and contemplation is definitely not feelings plus thoughts: it is the absence of both in formless intuition, which, being formless, can easily integrate the forms of Nature and of Mind, something that either or both together could never do for themselves. This was Schelling's great insight about the formless and the "indifference," the great Abyss or Emptiness from which Mind and Nature both issue, an Abyss alone that can ultimately heal. And this is why the Idealists sharply criticized the Romantics as being hopelessly lost and confused regarding this integration.)

THE GLORY OF THE VISION

This, truly, was a stunning vision, the likes of which humankind has rarely seen: evolution as Spirit's temporal unfolding of its own timeless potentials. Grounded in the pragmatic facts and actual history of consciousness, yet at the same time wedded to an all-pervading spiritual reality glorious in its grace and grand in its splendor, this Idealist vision brought Heaven down to awaken the Earth and brought Earth up to exalt its Heaven.

Idealism came very close to integrating the Big Three. There was abundant room for art, morals, and science, and they were carefully seen as important and cherished moments in the overall process of Spirit itself. Moreover, the Idealist vision was alive to the currents of development (or evolution). It was the first philosophy ever to come to terms with—and fully embrace—the sweeping implications of all-encompassing *development*, especially in religion and spirituality. Moreover, Idealism integrated Spirit and evolution in perhaps the only convincing way, namely, by recognizing that evolution is simply Spirit-in-action, or "God in the making."

Thus evolution, far from being an antispiritual movement—as so many Romantics and antimodernists and virtually all premodern cultures imagined—is actually the concrete unfolding, holarchical integration, and self-actualization of Spirit itself. Evolution is the mode and manner of Spirit's creation of the entire manifest world, not one item of which is left untouched by its all-encompassing embrace.

Henceforth, any spirituality that did *not* embrace evolution was doomed to extinction. Modern science, after the collapse, would reject the spiritual nature of evolution but retain the notion of evolution itself. Modern science, that is, would give us the exteriors of evolution—its surfaces and forms—but not its interiors—including Spirit itself. But even science would realize that evolution is universal, touching everything in existence, and, as Daniel Dennett put it, "like 'universal acid,' evolution eats through every other explanation for life, mind, and culture." How could it not, when it is actually Spirit-in-action, and Spirit embraces all?

Even though modern science has rejected the interiors of evolution while retaining the exterior surfaces, nonetheless science has amassed so much evidence for the existence of evolution in general that, to this day, any religion that attempts to reject evolution seals its own fate in the modern world. Even Pope John Paul II finally conceded that "evolution is more than a hypothesis."

One of the crucial ingredients in any integration of science and religion is the integration of empirical evolution with transcendental Spirit. The Idealists hit upon what very well might be the only conceivable way that this particular requirement can be met, namely, by seeing evolution as Spirit-in-action, thus accounting not only for the *what* and *when* of evolution (the empirical forms and Right-Hand surfaces accepted by modern science) but the *why* and *how* as well (the Left-Hand depths and interior intentionality of Spirit-in-action).

This extraordinary insight is to Idealism's everlasting credit. This lustrous vision saw the entire universe—atoms to cells to organisms to societies, cultures, minds, and souls—as the radiant unfolding of a luminous Spirit, bright and brilliant in its way, never-ending in its liberating grace. For, as Hegel put it, "Everything that from eternity has happened in heaven and earth, the life of God and all the deeds of time simply are the struggles for Spirit to know itself, to find itself, be for itself, and finally unite itself to itself; it is alienated and divided, but only so as to be able thus to find itself and return to itself." Involution is the story of that alienation, and evolution is the story of that extraordinary return.

THE LIMITATIONS OF IDEALISM

And yet, and yet. . . . There was at least one crippling inadequacy in Idealism, along with one major and devastating current in the modern world, that together brought Idealism tumbling down (although many of its core insights remain quite valid).

The inadequacy was that it possessed no yoga—that is, no tried and tested practice for *reliably reproducing* the transpersonal and superconscious insights that formed the very core of the great Idealist vision. Either these insights came spontaneously (and thus could not easily be reproduced), or they were the result of interior injunctions that were not anchored in dependable and *sustained* practice (and thus could not easily be reproduced).

Fichte, for example, used to perform this interior experiment with his students: "Be aware of the wall. Now be aware of that which is aware of the wall. Now be aware of that which is aware of that which is aware. . . ." In other words, this was a genuine if somewhat clumsy attempt to push back to the pure Witness, the absolute subjectivity that can never be seen as an object because it is the pure and formless Seer. Fichte wanted his students to contact what he called "the absolute Self," and you can begin to do so by inquiring within, asking "Who am I?" or "What is it that is now aware?" This radical Self, according to Fichte, is the source of the entire manifest world.

This, of course, is virtually identical to the great Vedanta notion of the identity of Atman (the pure Self in the individual) and Brahman (the Self of the Kosmos). Similar types of interior experiments were used by Vedanta, among others, to contact this pure Witness, with one major exception: these interior experiments or injunctions—known as yoga—were what we might call industrial-strength. They were not simply short exercises performed in the classroom to give students a glimpse of the divine Self; they were intense practices often pursued for hours, days, months, even years at a time, in virtually unbroken practice sessions.

In Zen, for example, if the koan (or meditation theme) is "Who am I?" or "Who chants the name of the Buddha?," it takes an average of *six years*, according to Yasutani Roshi, for the successful student to have the first profound satori, or genuine breakthrough to the True Self (which is the True

World as well). It is through that *sustained and intense practice* that actual transpersonal awareness of nondual Spirit is awakened, deepened, sustained, and transmitted from master to student.

The Idealists had none of this profound and sustained spiritual practice or yoga. Thus their transpersonal insights, profound as they were, came haphazardly and randomly; worse, they had no means of reliably reproducing these insights in others. Either you stumbled onto this transpersonal and superconscious experience, or you did not. If you did, you found that the Idealists spoke directly to you; if you did not, you found them completely confused and lost in metaphysical rubbish.

Lacking a genuine means of reproducible injunctions (or yoga), the Idealists' "transpersonal knowledge" was thus dismissed as "mere metaphysics," which, in the wake of Kant, was enough to doom any philosophy. And in a sense, precisely because the Idealists lacked a genuine spiritual injunction (practice, exemplar, paradigm), they were indeed, at least in this respect, caught in "mere metaphysics." For metaphysics in the "bad" sense is *any thought system without means of verification* (without validity claims or means of gathering actual data and evidence). Lacking the means of reproducibly generating actual and direct experiential evidence—*lacking the means of consistently delivering direct spiritual experience*—Idealism in this regard degenerated into abstract speculations without the means of experiential confirmation or rejection.

Thus, within a few decades of Hegel's death, the word was out: the Idealists did not in fact deliver the long-sought integration of the Big Three. They *talked* about it, but they did not seem to be able to *actually deliver* the experiential goods for other people. The modern dissociation had not been healed—if anything, it continued to accelerate—and the Idealists had been powerless to stop it.

In less than a century, the great Idealist vision had, for all practical purposes, come and gone. This glorious spiritual flower, perhaps the finest the modern West has ever known, saw its petals wither and fall, blown carelessly across a landscape increasingly flat and faded, the bleak and brave new world of the coming wasteland.

THE REIGN OF THE IT

The major interior deficiency of Idealism was the lack of a genuine yoga; the major exterior current in the modern (and soon-to-be postmodern) world that contributed to the devastation of the Idealist vision was simply the continuing collapse of the Kosmos.

Under the reign of it-science (quickly moving into its most powerful and imperial form as systems science, which saw the world as a holistic web of interwoven ITs) combined with it-industrialization (which objectified and commodified all human and intersubjective exchange, turning "I" and "we" into commercial "its" to be bought and sold in the marketplace)—under those combined forces, the Left-Hand and interior dimensions were being rapidly colonialized and enslaved by the aggressive Right-Hand domains. The value spheres of art and morals and spirituality, of interior consciousness and introspection and contemplation, of meaning and value and depth—in short, the Big Three—were rudely collapsed into the Big One of material monism.

We thus arrive at the official modern Western worldview—namely, *flatland holism:* atoms are parts of molecules which are parts of cells which are parts of organisms which are parts of societies of organisms which are parts of the biosphere which is part of the cosmos at large. However true the elements of that holarchy might be, they all have simple location and thus they all, without exception, are described in it-language and known in an empirical fashion.

This *subtle reductionism* simply reduces every holon in the Left Hand to its corresponding aspects in the Right Hand, thus gutting the interior dimensions and reducing them to empirical systems of ITs. This Right-Hand or flatland holism is a marvelously interwoven and coherent system. It acknowledges holarchies and systems and interwoven processes; it allows the brain and the organism and wonderfully complex ecosystems; it sees relationship upon relationship in never-ending process, all united in the wonderful Web of Life. It simply lacks, in irreducible terms, any actual consciousness, awareness, intentionality, feeling, introspection, contemplation, intuition, value, poetry, meaning, depth or Divinity.

Disenchanted, in other words, was fast becoming disemboweled. And against this flatland holism of scientific materialism, which both Romanticism and Idealism had failed to curb, came the first specifically postmodern revolts, in the more narrow and technical sense of postmodern poststructuralism. Since science arrogantly refused to take its place in a graceful integration with the other equally important value spheres, then let us simply crucify science, deconstruct science, right at its very foundations.

Having slain the Goliath of science, David and his fellow poets, artists, literary theorists, new-paradigm thinkers, and visionaries of every variety could now run free on the gloriously open field. A new age, surely, was about to dawn.

9

POSTMODERNISM:
TO DECONSTRUCT THE WORLD

If we use "postmodern" in the broad sense of any development occurring in the wake of modernity, then both Romanticism and Idealism can be taken as the first great
postmodern revolts against the dissociations and disasters of
flatland modernity. But with the continuing collapse of the
Kosmos—the denial of any substantive reality to any interior
or Left-Hand domain—neither Romanticism nor Idealism
could breathe; they slowly, inexorably suffocated to death,
and by the end of the nineteenth century they were basically
ineffectual as any sort of widespread cultural movements
with a *serious* chance of challenging scientific monism and
flatland holism.

And thus, from *within* the collapsed and postmodern Kosmos, there arose the first great attempt to unseat science, not
by arguing for higher modes of knowing (as both Romanticism and Idealism had done), but by *attempting to undermine
science in its own foundations.* There arose, that is, postmodernism in the narrow and specific sense (postmodern poststructuralism), generally associated with a list of names
stretching from Nietzsche to Heidegger to Bataille, Foucault,
Lacan, Deleuze, Derrida, Lyotard, and company (with a dose
of late Wittgenstein thrown in for good measure).

There is no way to understand postmodernism without
grasping the intrinsic role that *interpretation* plays in human
understanding. Postmodernism, in fact, may be credited with

making interpretation central to both epistemology and ontology, to both knowing and being. Interpretation, the postmodernists all maintained in their own ways, is not only
crucial for understanding the Kosmos, it is an aspect of its
very structure. *Interpretation is an intrinsic feature of the fabric
of the universe;* and there, in a sentence, is the enduring truth
at the heart of the great postmodern movements.

WHAT DOES THAT MEAN?

Many people are initially confused as to why, and how, interpretation is intrinsic to the universe. Interpretation is for
things like language and literature, right? Yes, but language
and literature are just the tip of the iceberg, an iceberg that
extends to the very depths of the Kosmos itself. We might
explain it like this:

All Right-Hand events—all sensorimotor objects and empirical processes and ITs—can be seen with the monological
gaze, with the eye of flesh. You simply look at the rock, the
town, the clouds, the mountain, the railroad tracks, the airplane, the flower, the car, the tree. All these Right-Hand objects and ITs can be *seen* by the senses or their extensions
(microscopes to telescopes). They all have simple location;
you can actually point to most of them.

But Left-Hand or interior holons cannot be seen in that
fashion. You cannot see love, envy, wonder, compassion, insight, intentionality, value, or meaning running around out
there in the empirical world. Interior events are not seen in
an *exterior* or *objective* manner, they are seen by *introspection*
and *interpretation*. Not merely the eye of flesh, but the eye of
mind (not to mention the eye of contemplation).

Thus, if you want to study *Macbeth* empirically, you can
get a copy of the play and subject it to various scientific tests:
it weighs so many grams, it has so many molecules of ink, it
has so many pages composed of such-and-such organic com-
pounds, and so on. That is all you can know about *Macbeth*
empirically. Those are its Right-Hand, objective, exterior aspects.

But if you want to know the *meaning* of the play, you will have to read it and enter into its interiority, its meaning, its intentions, its depths. The only way you can do that is by *interpretation:* What does this sentence *mean?* Here empirical science is virtually worthless, because we are entering interior domains and symbolic depths, which cannot be accessed by exterior empiricism but only by introspection and interpretation. Not just objective, but intersubjective. Not just monological, but dialogical.

Thus, you might see me coming down the street, a frown on my face. You can see that. But what does that exterior frown actually mean? How will you find out? You will ask me. You will talk to me. You can see my surfaces, but in order to understand my interior, my depths, you will have to enter into the interpretive circle. You, as a subject, will not merely stare at me as an object (of the monological gaze); rather, you, as a subject, will attempt to understand me, as a subject—as a person, as a self, as a bearer of intentionality and meaning. You will talk to me, and interpret what I say; and I will do the same with you. We are not subjects staring at objects; we are subjects trying to understand subjects—we are in the intersubjective circle, the dialogical dance. Monological is to describe; dialogical is to understand.

This is true not only for humans, but for all sentient beings as such. If you want to understand your dog—is he happy, or perhaps hungry, or wants to go for a walk?—you will have to *interpret* the signals he is giving you. And your dog, to the extent that he can, does the same with you. In other words, the *interior* of a holon can *only* be accessed by interpretation.

Thus, to put it bluntly, exterior surfaces can be *seen*, but interior depth must be *interpreted*. And precisely because this depth is an intrinsic part of the Kosmos—it is the Left-Hand dimension of every holon—interpretation itself is an intrinsic feature of the Kosmos. Interpretation is not something added onto the Kosmos as an afterthought; it is the very opening of the interiors themselves. And since the depth of the Kosmos goes "all the way down," then, as Heidegger famously put it, "Interpretation goes all the way down."

Perhaps we can now see why one of the great and noble

aims of postmodernism was to *introduce interpretation as an intrinsic aspect of the Kosmos.* As I would put it, every holon has a Left- and a Right-Hand dimension, and therefore every holon without exception has an objective (Right) and an interpretive (Left) component.

The disaster of modernity was that it reduced all introspective and interpretive knowledge to exterior and empirical flatland: it attempted to erase the richness of interpretation from the script of the world. (In postmodernese: Modernity marginalized the multivalent epistemic modes via an aggressive hegemony of the myth of the given that hierarchically inverted hermeneutic inscriptions due to the phallologocentrism of patriarchal signifiers. Translation: it collapsed Left to Right.)

Perhaps we can begin to see that the attempt by postmodernism to reintroduce interpretation into the very structure and fabric of the Kosmos was yet another attempt to escape flatland, to resurrect the gutted interiors and interpretive modes of knowing. The postmodern emphasis on interpretation—starting most notably with Nietzsche and running through Dilthey's *Geist* sciences to Heidegger's hermeneutic ontology to Derrida's "There is nothing outside the text [interpretation]"—is at bottom nothing but the Left-Hand domains screaming to be released from the crushing oblivion of the monological gaze of scientific monism and flatland holism. It was the bold reassertion of the I and the WE in the face of faceless ITs.

EXTREME POSTMODERNISM

Yet, as is so often the case with postmodernism, this moment of truth—every actual occasion has an interpretive component—was taken to absurd and self-defeating extremes: There is *nothing but* interpretation, and thus we can *dispense with the objective component of truth altogether* (in which case this theory cannot itself claim to be true: "So, if true, it is false. So, it is false." This, as we saw, is the performative contradiction hidden in all extreme postmodern "theoreticism," at which point this approach often hooks up with a mis-Kuhnian "new-paradigm" maneuver).

This extreme denial of any sort of objective truth amounts to a *denial of the Right-Hand quadrants altogether*, precisely the *reverse disaster* of modernity: all Right-Hand objects reduced to Left-Hand interpretations. And thus, all truth reduced to interpretive whim. Yet supposedly this reverse disaster will relieve modernity of its fragmented madness.

Since modern science had, in effect, killed two of the three value spheres (I-aesthetics and we-morals), postmodernism would simply attempt *to kill science as well*, and thus, in its own bizarre fashion, attempt an "integration" or "equal valuing" of all three spheres because all three of them were now equally dead, so to speak. Three walking corpses would heal the dissociations of modernity. Into the postmodern wasteland walked the zombie squad, and the wonder of it all is that they managed to convince a fair number of academics that this was a viable solution to modernity's ills.

Nonetheless, (extreme) postmodernism is now by far the most prevalent mood of academia, literary theory, the new historicism, a great deal of political theory, and (whether their proponents realize it or not) virtually all of the "new-paradigm" approaches to integrating science and religion. It thus behooves us to understand both its important truths and its extremist distortions.

MOMENTS OF TRUTH IN POSTMODERNISM

Postmodern philosophy is a complex cluster of notions that are defined almost entirely by what its proponents *reject*. They reject foundationalism, essentialism, and transcendentalism. They reject rationality, truth as correspondence, and representational knowledge. They reject grand narratives, metanarratives, and big pictures of any variety. They reject realism, final vocabularies, and canonical description.

Incoherent as the postmodern theories often sound (and

often are), most of these "rejections" stem from three core assumptions:

1. Reality is not in all ways pregiven, but in some significant ways is a construction, an interpretation (this view is often called "constructivism"); the belief that reality is simply given, and not also partly constructed, is referred to as "the myth of the given."
2. Meaning is context-dependent, and contexts are boundless (this is often called "contextualism").
3. Cognition must therefore privilege no single perspective (this is called "integral-aperspectival").

I believe all three of those postmodern assumptions are quite accurate (and need to be honored and incorporated in any integral view). Moreover, each tells us something very important with regard to any conceivable integration of science and religion, and thus they need to be studied with care. But each of those assumptions has also been blown radically out of proportion by the extremist wing of postmodernism, and the result is a totally deconstructed world that takes the deconstructionists with it.

Let us review those important truths—and their extremist contortions—one at a time.

THE MYTH OF THE GIVEN

We have already seen that Kant provided convincing arguments that much of what we take to be innocently given to us by the senses is actually a construction of the mind. For example, we say that we can easily see that our fingers are different from one another. But where is that difference located? Can you actually point to it? Can you see it? You can see the individual fingers, but can you actually *see* the difference between them?

The fact is, "difference" is a mental concept that we superimpose on certain raw sensations. Nowhere in those sensations do we actually experience or see "difference"—we *construct* it, *impose* it, *interpret* it; we never actually *perceive* it.

In other words, much of what we take to be *perceptions* are actually *conceptions*, mental and not empirical.

Thus, when many empiricists demand sensory evidence, they are actually demanding mental interpretations without realizing it. The Idealists, recall, took this fact and moved in a very "mental" direction: *Everything* we see is the product of mind (but a supraindividual and transpersonal mind or spirit). The postmodern poststructuralists took this notion and moved in a similar but much less spiritual and much more chaotic direction: The world given to us is not a perception but an interpretation, and thus there are no foundations, spiritual or otherwise, to ground anything.

It is exactly at that point that much of postmodernism starts to go extreme. It does not just emphasize the Left-Hand (or interpretive) aspects of all holons, *it attempts to completely deny reality to the Right-Hand (or objective) facets.* The important features of the Kosmos that are interpretive are made the *only* features in existence. Objective truth itself disappears into arbitrary interpretations, themselves imposed by power, gender, race, ideology, anthropocentrism, androcentrism, speciesism, imperialism, logocentrism, phallocentrism, phallologocentrism, or other varieties of utter unpleasantness (except for the claims of the theoreticists themselves, which are miraculously exempted from the charges of prejudice that are supposedly present in all claims—the performative contradiction).

But the fact that all holons have an interpretive as well as objective component does *not* deny the objective component, it merely situates it. Even Wilfrid Sellars, generally regarded as the most persuasive opponent of "the myth of the given"—the myth of direct realism and naïve empiricism, the myth that reality is simply given to us—maintains that, even though the manifest image of an object is in part a mental construction, it is *guided* in important ways by *intrinsic features* of sense experience, which is exactly why, as Kuhn knew, science can make *real* progress.

Thus, all Right-Hand exteriors, even if we superimpose conceptions upon them, nonetheless have various intrinsic features that are registered by the senses or their extensions, and in that general sense, all Right-Hand holons have some sort of objective reality. The "difference" between your fingers might be a mental construct, but the fingers themselves in some sense preexist your conceptualization of them; they are not totally or merely a product of mental constructions (which is exactly why a dog, a preconceptual infant, and a camera—all lacking a conceptual mind to do any constructing—will still register them). A diamond will cut a piece of glass, no matter what cultural words or concepts we use for "diamond," "cut," and "glass," and no amount of cultural constructivism will change that simple objective fact.

So it is one thing to point out the partial but crucial role that interpretation plays in our perception of the world (so that we can indeed deny the myth of the given). But to go to extremes and deny any moment of objective truth at all (and any form of correspondence theory or serviceable representation) is simply to render the discussion unintelligible.

No wonder John Searle had to beat this approach back in his wonderful book *The Construction of Social Reality*—as opposed to "the social construction of reality"—the idea being that cultural realities are constructed on a base of correspondence truth that grounds the construction itself, without which no construction could get under way at all. Once again, we can accept the partial truths of postmodernism—interpretation and constructivism are crucial ingredients of the Kosmos, all the way down—without going overboard and attempting to reduce all other quadrants and all other truths to that partial glimpse.

MEANING IS CONTEXT-DEPENDENT

The same caution applies to the second important truth of postmodernism, namely, that meaning is context-dependent. The word "bark," for example, means something entirely different in the phrases "the bark of a dog" and "the bark of a tree"—in other words, meaning is in many important ways dependent upon the context in which it finds itself. Moreover, these contexts are in principle *endless* or *boundless*, and thus there is no way finally to master and control meaning once and for all (because one can always imagine a further context that would alter the present meaning).

As I would put it, contexts are indeed boundless precisely because reality is composed of holons within holons within holons *indefinitely*, with no discernible bottom or top. Even the entire present universe is simply a part of the next moment's universe. Every whole is always a part, endlessly. And therefore every conceivable context is boundless. To say that the Kosmos is holonic is to say it is contextual, all the way up and down.

But that postmodern moment of truth has, once again, been deformed and pressed into self-contradictory duty by extreme postmodernists (particularly the branch known as "deconstruction," and especially its American proponents), who use it to deny that any sort of meaning actually exists or can be conveyed at all. Anytime science or traditional philosophy attempts to make a statement about the objective world, deconstruction will simply find a context that renders the statement absurd or self-contradictory, thus "deconstructing" the attempt. Since such a context can *always* be found (they are limitless), any and all meaning can be aggressively exploded and deconstructed right at the start. No wonder even Foucault referred to this extreme postmodernism as "terrorism." (Critics noted that these terrorists did not, however, attempt to deconstruct the meaning of "tenure," "pay raise," "promotion," or "salary"; these, apparently, are all very meaningful.)

But again, if that is a meaningful theory, its own meaning is meaningless. If it is so, then it isn't; so, it isn't. Contextualism, yes; extreme contextualism, no.

THE LINGUISTIC TURN

The importance of contextualism, interpretation, and hermeneutics in general came to the fore historically with what has been called *the linguistic turn* in philosophy—the general realization that language is not simply a representation of a pregiven world, but has a hand in the creation and construction of that world. With the linguistic turn, which began roughly in the nineteenth century, philosophers stopped using language to describe the world, and instead started looking at language itself.

Suddenly, language was no longer a simple and trusted tool. Metaphysics in general was replaced with linguistic analysis, because it was becoming increasingly obvious that language is not a clear window through which we innocently look at a given world; it is more like a slide projector throwing images against the screen of what we finally see. Language helps to create my world, and, as Wittgenstein would put it, the limits of my language are the limits of my world.

In many ways, "the linguistic turn" is just another name for the great transition from modernity to postmodernity. Where both premodern and modern cultures simply and naïvely used their language to approach the world, the postmodern mind spun on its heels and began to look at language itself. In the entire history of human beings, this, more or less, had never happened before. Some altogether startling findings were to result.

If we are to integrate the wisdom of yesterday with the knowledge of today—and that means, in the broadest sweep, the best of premodern, modern, and postmodern—we will have to look carefully at what the postmodern linguistic turn brought to our understanding of the Kosmos. For the integration of science and religion is a camel that, one way or another, must be able to pass through the eye of the postmodern needle: constructivism, contextualism, and integral-aperspectival—all of which came to the fore with the linguistic turn.

LANGUAGE SPEAKS

Most forms of postmodern poststructuralism trace their lineage to the work of the brilliant and pioneering linguist Ferdinand de Saussure. Saussure's work, and especially his *Course in General Linguistics* (1916), was the basis of much of modern linguistics, semiology (semiotics), structuralism, and hence poststructuralism, and his essential insights are as cogent today as they were when he first advanced them almost a century ago.

According to Saussure, a linguistic *sign* is composed of a material *signifier* (the written word, the spoken word, the marks on this page) and a conceptual *signified* (what comes to mind when you see the signifier), both of which are different from the actual *referent*. For example, if you see a tree, the actual tree is the referent; the written word "tree" is the signifier; and what comes to mind (the image, the thought, the mental picture or concept) when you read the word "tree" is the signified. The signifier and the signified together constitute the overall sign.

But what is it, Saussure asked, that allows a sign to mean something, to actually *carry meaning*? For example, bark of a dog, bark of a tree. As we saw, the word "bark" has meaning, in each case, because of its place in the phrase (a different phrase gives the same word a totally different meaning). Each phrase likewise has meaning because of its place in the larger sentence and, eventually, in the total linguistic structure. Any given word in itself is basically *meaningless* because the same word can have different meanings depending on the context or the structure in which it is placed.

Thus, Saussure pointed out, it is the *relationship among all of the words themselves* that stabilizes meaning (and not merely some simple pointing to an object, because that pointing cannot even be communicated without a total structure that holds each word in meaningful place). So— and this was Saussure's great insight—*a meaningless element becomes meaningful only by virtue of the total structure.* (This is the beginning of *structuralism*, virtually all schools of which trace their lineage in whole or part to Saussure. Present-day descendants include aspects of the work of Lévi-Strauss, Jakobson, Piaget, Lacan, Barthes, Foucault, Derrida, Habermas, Loevinger, Kohlberg, Gilligan . . . it was a truly stunning discovery.)

In other words—and no surprise—every sign is a holon, a context within contexts within contexts in the overall network. And this means, said Saussure, that the entire language is instrumental in conferring meaning on an individual word.

Now, the standard Enlightenment (and flatland) notion was that a word gains meaning simply because it *points to* or *represents* an object. It is a purely monological and empirical affair. The isolated subject looks at an equally isolated object (such as a tree), and then simply chooses a word to represent the sensory object. *This, it was thought, is the basis of all genuine knowledge.* Even with complex scientific theories, each theory is simply a *map* that *represents* the objective territory. If the correspondence is accurate, the map is true; if the correspondence is inaccurate, the map is false. Science—and all true knowledge, it was believed—is a straightforward case of *accurate representation*, accurate mapmaking. "We make pictures of the empirical world," as Wittgenstein would soon put it, and if the pictures match, we have the truth.

This is the so-called *representation paradigm*, which is also known as the *fundamental Enlightenment paradigm*, because it was the general theory of knowledge shared by most of the influential philosophers of the Enlightenment, and thus modernity in general. (Recall that in Chapter 4 I actually listed that as one of the defining aspects of modernity: "Modern philosophy is usually 'representational,' which means it tries to form a correct representation of the world. This representational view is also called 'the mirror of nature,' because it was commonly believed that the ultimate reality was sensory nature and philosophy's job was to picture or mirror this reality correctly.")

It was not the existence or the usefulness of representation that was the problem; representational knowledge is a perfectly appropriate form of knowing for many purposes. Rather, it was the aggressive and violent attempt to reduce all knowledge to empirical representation that constituted the disaster of modernity—the reduction of translogical spirit and dialogical mind to monological sensory knowing: the collapse of the Kosmos to nothing but Right-Hand representation.

Saussure, with his early structuralism, gave one of the first, and still one of the most accurate and devastating, critiques of empirical theories of knowing, which, he pointed out, cannot even account for the simple case of "the bark of a tree." The meaning comes not merely from *objective* pointing but from *intersubjective* structures that *cannot themselves be objectively pointed to!* Yet without them, there would, and could, be no objective representation at all!

So what I, as a proper Enlightenment philosopher, took to be a simple "representation" is not so simple after all. I thought that I, the autonomous subject, the isolated and independent self, could simply choose a word (such as "bark") and then say what object I wanted it to point to, to represent. So I imagine that I am utterly prior to this creation of meaning—I am the proud and autonomous subject that creates all this meaning by simply pointing to the objects that I mean.

The reality is pretty much the opposite: Meaning is created for me by vast networks of background contexts about which I consciously know very little. I do not fashion this meaning; this meaning fashions me. I am part of a vast background of cultural signs, and in many cases I have no clue as to where it all came from.

In other words, every subjective intentionality (Upper Left) is *situated* in vast networks of intersubjective or cultural contexts (Lower Left) that are instrumental in the creation and interpretation of meaning itself. Meaning is not merely *objective* pointing but *intersubjective* networks; not simply *monological* but *dialogical;* not just *empirical* but *structural;* not just *representational* images but systemic *networks*—and the meaning is as much a result of the network as of the referent. This is precisely why meaning is indeed context-dependent, and why the bark of a dog is different from the bark of a tree.

In the wake of this extraordinary linguistic turn, philosophers would never again look at language in a simple, trusting way. Language does not merely report the world, represent the world, describe the world. Rather, language creates worlds, and in that creation is power. Language creates, distorts, carries, discloses, hides, allows, oppresses, enriches, enthralls. For good or ill, language itself is something of a demigod, and philosophers henceforth would focus much of their attention on that powerful force. From linguistic analysis to language games, from structuralism to poststructuralism, from semiology to semiotics, from linguistic intentionality to speech act theory, postmodern philosophy has been in large measure *the philosophy of language*, and it pointed out—quite rightly—that if we are to use language as a tool to understand reality, we had better start by looking very closely at that tool.

LANGUAGE GROANS

The postmodern poststructuralists took many of these profound and indispensable notions and, in carrying them to extremes, rendered them virtually useless. They did not just *situate* individual intentionality in background cultural contexts, they tried to *erase* the individual subject altogether: "the death of man," "the death of the author," "the death of the subject"—all were naked attempts to reduce the subject (Upper Left) to nothing but intersubjective structures (Lower Left). "Language" replaced "humans" as the *agent* of history. It is not I, the subject, who is now speaking, it is nothing but impersonal language and linguistic structure speaking through me.

Thus, as only one of innumerable examples, Foucault would proclaim that "Lacan's importance comes from the fact that he showed how it is the structures, the very system of language, that speak through the patient's discourse and the symptoms of his neurosis—not the subject." Upper Left reduced to Lower Left, to what Foucault famously called "this anonymous system without a subject."

Thus I, Ken Wilber, am not writing these words, nor am I in any way primarily responsible for them; language is actually doing all the work (although this did not prevent I, Roland Barthes, or I, Michel Foucault, from accepting the royalty checks written to the author that supposedly did not exist).

Put simply, the fact that each "I" is always situated in a background "We" was perverted into the notion that there is no "I" at all, but only an all-pervading "We"—no individual subjects, only vast networks of intersubjective and linguistic structures. (Buddhists, take note: this was in no way the notion of *anatta*, or no-self, because the "I" was replaced, not with Emptiness, but with finite linguistic structures of the "We," thus multiplying, not transcending, the actual problem.)

Foucault eventually rejected the extremism of his early stance, a fact studiously ignored by extreme postmodernists. Among other hilarious spectacles, postmodernist biographers began trying to write biographies of subjects that supposedly did not exist in the first place, thus producing books that were about as interesting as dinner without food.

For Saussure, the signifier and signified were an integrated unit (a holon); but the postmodern poststructuralists—and this was one of their most defining moves—shattered this unity by attempting to place almost exclusive emphasis on sliding chains of *signifiers* alone. The signifiers—the actual material or written marks—were given virtually exclusive priority. They were thus severed from both their signifieds and their referents, and these chains of sliding or "free-floating" signifiers were therefore said to be anchored in nothing but power, prejudice, or ideology. (We see again the extreme constructivism so characteristic of postmodernism: signifiers are not anchored in any truth or reality outside of themselves, but simply create or construct all realities.)

Sliding chains of signifiers: this is the essential postmodern poststructuralist move. It is postSTRUCTURAL, because it starts with Saussure's insights into the networklike structure of linguistic signs, which partially construct as well as partially represent; but POSTstructural, because the signifiers are cut loose from any sort of anchoring at all. There is no objective truth (only interpretations), and thus, according to extreme postmodernists, signifiers are grounded in nothing but power, prejudice, ideology, gender, race, colonialism, speciesism, and so on (a performative contradiction that would mean that this theory itself must also be anchored in nothing but power, prejudice, etc., in which case it is just as vile as the theories it despises).

This is exactly where the postmodern agenda would often hook up with the mis-Kuhnian notion of "paradigm." It was a marriage made in interpretive heaven for all those who wished to "deconstruct" the "old paradigm" and replace it with the "new paradigm," which itself lacked any genuine exemplars or injunctions and thus, according to Kuhn's actual notion of paradigm, was no such thing at all, but merely ideology dressed up as cultural studies, narcissism and nihilism in transformational drag.

INTEGRAL-APERSPECTIVAL

The fact that meaning is context-dependent—the second important truth of postmodernism, also called "contextualism"—means that a multiperspective approach to reality is called for. Any single perspective is likely to be partial, limited, perhaps even distorted, and only by taking multiple perspectives and multiple contexts can the knowledge quest be fruitfully advanced. And that "diversity" is the third important truth of general postmodernism.

Jean Gebser, whom we have seen in connection with worldviews, coined the term *integral-aperspectival* to refer to this pluralistic or multiple-perspectives view, which I also refer to as *vision-logic* or *network-logic*. "Aperspectival" means that no single perspective is privileged, and thus, in order to gain a more holistic or *integral* view, we need an *aperspectival* approach, which is exactly why Gebser usually hyphenated them: integral-aperspectival.

Gebser contrasted integral-aperspectival cognition with formal rationality, or what he called "perspectival reason," which tends to take a single, monological perspective and view all of reality through that narrow lens. Where perspectival reason privileges the exclusive perspective of the particular subject, vision-logic *adds up all the perspectives*, privileging none, and thus attempts to grasp the integral, the whole, the multiple contexts within contexts that endlessly disclose the Kosmos, not in a rigid or absolutist fashion, but in a fluidly holonic and multidimensional tapestry.

This parallels almost exactly the Idealists' great emphasis on the difference between a reason that is merely monological, representational, or empiric-analytic, and a reason that is dialogical, dialectical, and network-oriented (vision-logic). They called the former *Verstand* and the latter *Vernunft*. And they saw *Vernunft* or vision-logic as being a higher evolutionary development than mere *Verstand* or formal rationality. In fact, they tended to view monological or perspectival rationality as a "monster of arrested development"—and that monological monster was, of course, the mode of knowing that largely defined the Enlightenment, which is why the Idealists' critique of the Enlightenment (and flatland modernity) is still one of the most powerful and cogent ever advanced.

Gebser, too, believed that vision-logic was an evolutionary development beyond monological rationality. Nor are Gebser and the Idealists alone. Many schools of transpersonal psychology and sociology, not to mention important conventional theorists from Jürgen Habermas to Carol Gilligan, see dialectical vision-logic as a higher and more embracing mode of reason. (This is shown in Figure 5-1, where "formop" is formal rationality and "vision-logic" is integral-aperspectival. Vision-logic is not yet transrational but, we might say, lies on the border between the rational and the transrational, and thus partakes of some of the best of both.)

This vision-logic not only can spot massive interrelationships, it is itself an intrinsic part of the interrelated Kosmos, which is why vision-logic does not just *represent* the Kosmos, but is a *performance of* the Kosmos. Of course, all modes of genuine knowing are such performances; but vision-logic is the first that can self-consciously realize this and articulate it. Hegel did so in the first extensive and pioneering fashion—vision-logic evolutionarily became conscious of itself in Hegel—and Saussure did exactly the same thing with linguistics. Saussure took vision-logic and applied it to language, thus disclosing, for the first time in history, its network structure. The linguistic turn is, at bottom, vision-logic looking at language itself.

This same vision-logic would give rise to the extensively

elaborated versions of systems theory in the natural sciences; it would stand as well behind the postmodernists' recognition that meaning is context-dependent and contexts are boundless. In all these movements and more, we see the radiant hand of vision-logic announcing the endless networks of holonic interconnection that constitute the very fabric of the Kosmos itself.

This is why I believe that the recognition of the importance of integral-aperspectival cognition is the third great (and valid) message of postmodernism in general. This is likewise why one of the ways we can date the beginning of the general postmodern mood is with the great Idealists (note that Derrida does exactly that; Hegel, he says, is the last of the old or the first of the new).

LANGUAGE COLLAPSES

All of which is well and good. But it is not enough, we have seen, to be "holistic" instead of "atomistic," or to be network-oriented instead of analytic and divisive. Because the alarming fact is that *any mode of knowing can be collapsed* and confined merely to surfaces, to exteriors, to Right-Hand occasions. And, in fact, almost as soon as vision-logic had heroically emerged in evolution, it was crushed by the flatland madness sweeping the modern world.

Indeed, as we have repeatedly seen, the systems sciences themselves did exactly that. The systems sciences denied any substantial reality to the I and the WE domains (in their own terms), and reduced all of them to nothing but interwoven ITs in a dynamical system of network processes. This was vision-logic at work, but a crippled vision-logic, hobbled and chained to the bed of exterior processes and empirical ITs. This was a holism, but merely an exterior holism that perfectly gutted the interiors and denied any sort of validity to the extensive realms of Left-Hand holism (of the I and the WE). The shackles were no longer atomistic; the shackles— and don't we all feel better?—were now holistically interwoven chains of degradation.

Precisely the same fate awaited so much of the general postmodern agenda. Starting from the admirable reliance on vision-logic and integral-aperspectival awareness—yet still unable to escape the collapse of the Kosmos—these postmodern movements ended up subtly embodying and even extending the reductionistic nightmare. They all became a new and higher form of reason, yes, but *reason still trapped in flatland*. They were perfectly, but perfectly, another twist on flatland holism, material monism, monological madness. They still succumbed to the disaster of modernity even as they loudly announced they had overcome it, subverted it, inverted it, deconstructed it, exploded it.

There is nothing but *sliding chains of signifiers:* you see, the only reality is sliding chains *of material marks*—in other words, sliding chains of ITs. For all the emphasis on interpretation and interior validation, postmodern poststructuralism all comes down to sliding chains of material ITs. Gone are the actual signifieds—the actual interior domains of the I and the WE disclosed in their own terms—and what we have left, as with systems theory, are holistic chains of interwoven ITs, holistic surfaces, all utterly lacking any genuine depth whatsoever, and thus utterly incapable of curing the dissociations of modernity. And so it came about that extreme postmodernism was simply part of the insidious disease for which it loudly claimed to be the cure.

DEPTH TAKES A VACATION

In fact, most postmodernists would go to extraordinary lengths to deny depth in general. It is as if, suffering under the onslaught of flatland aggression, postmodernism identified with the aggressor. Postmodernism came to embrace surfaces, champion surfaces, glorify surfaces, and surfaces alone. There are only sliding chains of signifiers; everything is a material text; there is nothing under the surface; there is only the surface. As Bret Easton Ellis put it in *The Informers:* "Nothing was affirmative, the term 'generosity of spirit' applied to nothing, was a cliche, was some kind of bad joke. . . .

Reflection is useless, the world is senseless. Surface, surface, surface was all that anyone found meaningful . . . this was civilization as I saw it, colossal and jagged."

Robert Alter, reviewing William H. Gass's *The Tunnel*—a book claimed by many to be the ultimate postmodern novel—points out that the defining strategy of this postmodern masterpiece is that "everything is deliberately reduced to the flattest surface." It does so by "denying the possibility of making consequential distinctions between, or meaningful rankings of, moral or aesthetic values. There is no within: murderer and victim, lover and onanist, altruist and bigot, dissolve into the same ineluctable slime"—the same sliding chains of equally meaningless signifiers.

Everything reduced to the flattest surface. . . . *There is no within*—a perfect description of flatland, a flatland that, beginning with modernity, was actually amplified and glorified with extreme postmodernity: "surface, surface, surface was all that anyone found."

Alter is exactly right that behind it all is the inability or refusal to make "consequential distinctions between, or meaningful rankings of, moral or aesthetic values." In this wasteland where Right-Hand signifiers and surfaces alone exist, there are no value, no meaning, and no qualitative distinctions of any sort, for those exist only in the Left-Hand domains. To collapse the Kosmos to Right-Hand signifiers is to step out of the real world and into the twilight zone known as the disqualified universe. Here there are no interior holarchies, no meaningful rankings of the I and the WE, no qualitative distinctions of any sort and no gradations of depth, so that fact and fiction, truth and lies, murderer and victim, as Alter said, are all reduced to equivalent surfaces.

"Subvert all hierarchies!"—one of the battle cries of extreme postmodernism—actually means "Destroy all value, kill all quality, massacre all meaning." Extreme postmodernism went from the noble insight that all perspectives need to be given a fair hearing, to the utterly self-contradictory belief that no perspective whatsoever is better than any other (self-contradictory because their own belief is held to be much better than the alternatives).

Thus, under the intense gravity of flatland, integral-aperspectival awareness became simply *aperspectival madness*—the contradictory belief that no belief is better than any other—a total paralysis of thought, will, and action in the face of a million perspectives all given exactly the same depth, namely, zero.

At one point in *The Tunnel*, Gass himself, the author of this postmodern masterpiece, describes the *perfect postmodern form*, which serves "to raunchify, to suburp [*sic*] everything, to pollute the pollutants, explode the exploded, trash the trash. . . . It is all surface. . . . There's no inside however long or far you travel on it, no within, no deep."

No within, no deep. That may serve as a perfect credo for extreme postmodernism in general. And it is into this modern and postmodern wasteland—and against its dominant and domineering mood—that we wish to introduce the within, the deep, the interiors of the Kosmos, the contours of the Divine.

PART III

A RECONCILIATION

———— ❦ ————

10

THE WITHIN: A VIEW OF THE DEEP

The modern and postmodern world is still living in the grips of flatland, of surfaces, of exteriors devoid of interior anything: "no within, no deep." The only large-scale alternatives are an exuberant embrace of shallowness (as with extreme postmodernism), or a *regression* to the *interiors* of *premodern* modes, from mythic religion to tribal magic to narcissistic new age. A modern and postmodern spirituality has continued to elude us, primarily because the *irreversible* differentiations of modernity have placed difficult but unavoidable demands on the sought-after integration: spirituality must be able to stand up to scientific authority, not by aping the monological madness but by announcing its own means and modes, data and evidence, validities and verifications. Spirituality must be able to integrate the Big Three value spheres of self, culture, and nature, not merely attempt to dedifferentiate them in a premodern slide or deconstruct them in a postmodern blast.

We have seen that historically the three most important attempts to introduce Spirit into the modern and postmodern world were Romanticism, Idealism, and some schools of Postmodernism. And we saw why each ultimately failed. The Romantics, caught in the pre/trans fallacy, often ended up recommending dedifferentiation instead of genuine integration (no matter what lip service they paid to the latter). The Idealists avoided regression but had no yoga, no transpersonal injunctions to reproduce their spiritual intuitions in others (so that Idealism degenerated into "mere metaphysics," or thought without actual evidence). And the Postmodernists, unanchored in any conception of truth, had nothing left but their own dispositions: narcissism and nihilism as a postmodern tag team from hell. Coupled with a mis-Kuhnian notion of "paradigm," egocentrism rushed in to announce the dawn of the glorious world transformation, gleefully giddy at the thought of how central its own ego was to the entire global show.

With the dominance of flatland, two major types of attempts *to break the hold of scientific monism* arose: Romanticism and Idealism attempted a type of epistemological pluralism, while postmodernism opted for "theoreticism." In the former, science is seen as one of several valid modes of knowing, so that science and religion can coexist as different but equally important aspects of the Real. In the latter, all modes of knowing are deconstructed, leaving an *extremely* level playing field, which, it is hoped, will be liberating due to a subverting of the dominator hierarchies.

All of them ultimately failed. Epistemological pluralism had been embraced in various forms by classical religion, Romanticism, and Idealism. But none of those forms could withstand the assault of an aggressive modern scientific monism (or epidemic systems it-ism). Modern science simply discovered that too many items that were supposed to be completely transcendent actually had very real, very immanent, and very natural anchors in the empirical brain and organism—nothing "otherworldly" about them, and thus no other modes of knowing that needed to be acknowledged. Epistemological pluralism collapsed with the rest of the Kosmos: there are only Right-Hand occasions accessed by monological and empirical modes of knowing. Science alone rules, the epidemic reign of the omnipresent It.

Postmodern theoreticism (science as poetry/paradigm) also collapsed, this time under the weight of its own absurdities and performative self-contradictions. Science and religion were placed on equal footing by killing them both, and the wasteland of endless surfaces was proclaimed the only real, a proclamation that, if true, is false.

Nonetheless, each of those approaches—Romanticism, Idealism, Postmodernism—has moments of truth that need to be honored and incorporated in a more integrative embrace. This, indeed, is one of the aims of a truly Integral approach, which this book is attempting to express.

If a genuine integration of science and religion is to be a reality, it will have to include an integration of the Big Three (of art, morals, and science), not by *deforming* any of them to fit some sort of pet scheme, but by taking each of them more or less exactly as they are. There is no need to force science into some sort of "new paradigm" that will then supposedly be compatible with spirituality. The very attempt is a massive category error, a profound confusion about the nature and role of monological science, dialogical philosophy, and translogical spirituality. These are to be integrated as we find them, not as we deform them in a monological leveling that erases the very differences that are supposed to be integrated in the first place.

The Integral approach, therefore, attempts just that—an integration, just as they are, of the Big Three of art (aesthetic-expressive, self and self-expression, subjective phenomenology), morals (intersubjective justness, ethical goodness, cultural communion), and science (objective nature, the empirical world, concrete occasions). Nothing spectacular has to be done to any of these three value spheres (or four quadrants); we take them more or less as we find them. All that is required is that each begin to harbor the suspicion that its truth is not the only truth in the Kosmos.

Nonetheless, exactly there is the difficulty. All of the past forms of epistemological pluralism *failed the test of modernity* because science itself did not and would not fundamentally doubt its own competence to reveal all important forms of truth. With the collapse of the Kosmos, the integration of the Big Three was no longer even perceived to be a problem. It was not a problem because there was no Big Three, only the Big One of scientific materialism and flatland holism: no integration was needed, none was sought.

In order to proceed effectively on our quest, we therefore need to back up prior to the collapse of the Kosmos. Not prior to the differentiations of modernity, but merely prior to the dissociations—and begin our reconstruction at that fateful point, that point of fragmentation, alienation, separation, and collapse.

And that was the point that *any and all interior dimensions lost legitimacy.* Modernity did not reject Spirit per se. Modernity rejected *interiors* per se, and Spirit was simply one of the numerous casualties. The within itself—from the lowest to the highest, from prepersonal to personal to transpersonal—the interiors themselves were objectified, turned into objects of the monological gaze, forced under the instrumental scope of scientific materialism: subjective mind reduced to material brain, intersubjective values reduced to technical steering problems, intentionality reduced to behavioral conditioning, Spirit (if it survived at all) reduced to the empirical Web of Life and econature (flatland holism), cultural worldviews reduced to material modes of production, compassion reduced to serotonin, consciousness reduced to digital bits—in short, and in all ways, I and WE reduced to scurrying ITs.

Granted, Spirit did not survive this modern collapse. But the point is, neither did any other interior dimension, including the simplest feelings, perceptions, and affects—not to mention intentionality in general, the mental domains as irreducible realities with their own ontological weight, and, yes, the spiritual and transpersonal realms as well. Flatland accepts no interior domain whatsoever, and reintroducing Spirit is the least of our worries.

Thus our task is not specifically to reintroduce spirituality and somehow attempt to show that modern science is becoming compatible with God. That approach, which is taken by most of the integrative attempts, does not go nearly deep enough in diagnosing the disease, and thus, in my opinion, never really addresses the crucial issues.

Rather, it is the rehabilitation of the *interior in general* that opens the possibility of reconciling science and religion, integrating the Big Three, overcoming the dissociations and disasters of modernity, and fulfilling the brighter promises of postmodernity. Not Spirit, but *the within*, is the corpse we must first revive.

THE OBJECTIONS
OF EMPIRICAL SCIENCE

All of the typical attempts to integrate science and religion have consistently failed because empirical science rejects the interior dimensions. It does so for two major reasons:

1. The allegedly "interior," "higher," "transcendental," "otherworldly," or "mystical" modes of consciousness all seem to be thoroughly embedded in natural, objective, empirical processes in the brain. Thus they are not higher in any genuine sense, but merely different types of biomaterial events in the biomaterial brain. No higher levels of reality, beyond the sensorimotor, are needed to explain any of these states.

Thus there are no irreducible interior domains that can be studied by different modes of knowing, there are only objective ITs (atomistic or holistic) studied best by science. In short, interior domains have no reality of their own; thus there are no "interior" modes of knowing that cannot be explained away, literally.

2. Even if there were other modes of knowing than the sensory-empirical, they would have no means of validation and thus could not be taken seriously. They are, at best, merely personal or subjective tastes and idiosyncratic displays, useful perhaps as emotional preferences but with no cognitive validity at all.

I believe, of course, that both of those objections—there are no interiors, and even if there are, they cannot be verified—are profoundly incorrect. Yet they stand like a reinforced brick wall across the road to the marriage of science and religion. If you find my responses to these two objections satisfactory, fine; if not, perhaps they will help you think of better ones. But unless and until these two objections are directly met and countered, there will be precisely zero integration of science and religion, and approaches not directly addressing them are largely irrelevant.

Here is a brief overview of my responses, which I trust will come more alive as we proceed.

I believe that objection number 1 (there are no real interiors) can be answered by the compelling evidence for the existence of all four quadrants, which themselves have an enormous amount of data—empirical, phenomenological, cross-cultural, and contemplative—supporting their existence. The sometimes staggering weight of this evidence places what many would see as a frightful burden on would-be reductionists, who would have to work relentlessly to erase it all to their satisfaction.

More to the point, if empirical science rejects the validity of any and all forms of interior apprehension and knowledge, then it rejects its own validity as well, a great deal of which rests on interior structures and apprehensions that are *not* delivered by the senses or confirmable by the senses (such as logic and mathematics, to name only two).

If science acknowledges these interior apprehensions, upon which its own operations depend, then *it cannot object to interior knowledge per se*. It cannot toss *all* interiors into the garbage can without tossing itself with it. (As we will see, most of the philosophers of science have *already* conceded this point.) This undermines objection number 1, leaving science to resort solely to objection number 2 in order to disqualify the other modes of knowing and maintain its hegemony.

I will then argue that objection number 2 can be answered by showing that the scientific method, in general, consists of three basic strands of knowing (injunction, apprehension, confirmation/rejection). If it can be shown that *the genuine interior modes of knowing also follow these same three strands*, then objection number 2—that these alternative modes have no legitimate validity claims—would be substantially refuted.

With the two major scientific objections to the interior domains undone, the door would be open to a genuine reconciliation of science and religion (and the Big Three in general). We will attempt to demonstrate to science, *not* that it is wrong, old paradigm, a product of dissociated consciousness, patriarchal, divisive, or sick to the core, but rather that it is correct but partial. Science, quite rightly, will accept none of the former critiques; if it will budge at all, it will have to be here.

THE RESURRECTION OF
THE INTERIOR

As we have seen, empiricists (and positivists and behaviorists and scienticians in general) deny irreducible reality to virtually all Left-Hand dimensions; only the Right Hand is real. All Left-Hand occasions are at best reflections or representations of the sensorimotor world, the world of simple location, the world of ITs, detected by the human senses or their extensions (or, in general, by some sort of objectifying activity).

In other words, they all subscribe to the myth of the given, the myth that the sensorimotor world is simply given to us in direct experience and that science carefully and systematically reports what it there finds. But this view is indeed a myth, and even most orthodox philosophers of science now concede this elemental point, so much so that it is already regarded as something of a settled issue.

We already saw that the myth of the given has been blown out of proportion by extreme postmodernists and then used to deny any objective truth whatsoever. This is an extremism we certainly wish to avoid. There are indeed what Wilfrid Sellars calls *intrinsic features* of the sensorimotor world that prevent the total dissolution of objective truth and allow science to make *real* progress.

But by the same token, a naïve empiricism—that science simply and innocently reports to us the unshakable givens of experience—is likewise an extreme and untenable view. It is the myth of the given.

We do not, for example, perceive a tree. What we actually see, what is given in our experience, is simply a bunch of colored patches. On this, empiricists, rationalists, and Idealists all agree. The traditional empiricist then attempts to ground all knowledge in these sensory "givens"—*the colored patches.* But it is now widely acknowledged that you cannot derive knowledge solely from patches. Classical empiricism has run aground on just this impossibility.

Thus, even *The Cambridge Dictionary of Philosophy,* long a bastion of the orthodox and consensus view, soberly points out that "Epistemologies postulating [simple] givenness require a single entity-type to explain the sensorial nature of perception and to provide immediate epistemic foundations for empirical knowledge. *This requirement is now widely regarded as impossible to satisfy;* hence Wilfrid Sellars describes the discredited view as the myth of the given. . . . Concluding that the doctrine of the given is false, he maintains that classical empiricism is a myth" (my italics).

And not just Sellars. As the dictionary itself points out, the requirements of classical empiricism are "now widely regarded as impossible to satisfy . . . classical empiricism is a myth." This parallels nicely John Passmore's summary of the state of positivism, the official philosophy of scientism: "Logical positivism, then, is dead, or as dead as a philosophical movement ever becomes."

This can all be put fairly simply: science approaches the empirical world with a massive conceptual apparatus containing everything from tensor calculus to imaginary numbers to extensive intersubjective linguistic signs to differential equations—virtually all of which are *nonempirical* structures found *only* in interior spaces—and then it astonishingly claims it is simply "reporting" what it "finds" out there in the "given" world—when, in fact, all that is given is colored patches.

For science to acknowledge the massive interior structures that it brings to the party is *not* to deny the objective intrinsic features of the exterior world; it is simply to recognize as well the reality and importance of the subjective and intersubjective domains responsible for generating so much of the knowledge.

Thus there are indeed preexisting intrinsic features in the sensorimotor world that constrain our perceptions—for example, if you drop the colored patch called an apple, it always falls to the colored patch called the ground. These intrinsic features *anchor* the objective component of truth (in any domain).

At the same time, these objective features are differentiated, conceptualized, organized, and given much of their actual form and content by conceptual structures that themselves exist *in nonempirical and nonsensory spaces.* These *interior structures* include not only deeply background cultural contexts, intersubjective linguistic structures, and consensus ethical norms, but also most of the specific conceptual tools that scientists use as they analyze their objective data, tools such as logic, statistical displays, and all forms of mathematics, from algebra to Boolean algebra to calculus to complex numbers to imaginary numbers. *None of these structures can be seen or found anywhere in the exterior, empirical, sensory world.* They are all, all of them, subjective and intersubjective occasions, interior occasions, Left-Hand occasions. And nobody has ever found any way whatsoever to reduce this knowledge to colored patches.

TO INVESTIGATE THE INTERIORS

Of course, empirical science is free to go on its merry way without stopping to look at the interior tools it uses in its assault on the world. What it is not allowed to do, without self-obliteration, is deny the existence or the importance of these tools. Yet that is exactly what happens when science degenerates into scientism and rejects in toto the existence of the interior dimensions, simply because none of them is a colored patch.

Empirical science depends upon these interior domains (subjective and intersubjective) for its own objective operation. But because they *cannot be accessed* by simple monological and objective and sensorimotor methods, empirical science, in its more brutish forms, has simply rejected these interiors altogether, interiors which not only allow its own operations, but also contain the within of the Kosmos.

This self-obliterating reductionism is not genuine science, it is simply science the village idiot. And, as everybody knows, it takes a village to raise a complete idiot—the village of collapsed modernity, in this case. Science becomes imperial scientism and falls into the simpleminded myth of the given, naïvely ascribing to its colored patches a great deal of what is found only in its conceptual apparatus, whose existence it has just denied.

But the crucial point is that these interior spaces and structures—from linguistics to mathematics to interpretive modes to logic—*can be investigated in their own right.* Scientists already do this with logic and mathematics. Nobody has ever seen imaginary numbers (such as the square root of negative one) running around in the sensory world, but mathematical scientists use them all the time, and they do so by investigating *the interior structures and patterns* that string various, nonempirical symbols together. The same can be said for most forms of logic, *n*-dimensional theories, tensor calculus . . . the list is almost endless.

We already saw the same approach applied to linguistics, where Saussure, in a historically groundbreaking move, rejected the myth of the given (simple empirical representation) and demonstrated that the meaning of a word comes not simply from its pointing to a colored patch, but also from its being part of a vast intersubjective network of nonempirical signs (none of which are merely colored patches).

Behaviorist theories of language can investigate only the simple pointing (and thus, as Chomsky notes, have never been able to explain language acquisition at all!). But *semiotics* (the study of signs in their intersubjective settings) and *hermeneutics* (the study of interpretation based on grasping the entire network of meaning) have made stunning advances in our understanding of linguistics, precisely by denying the myth of the given, the myth that the monological sensory world alone is the sole irreducible reality.

The conclusion is straightforward: the interior spaces not only structure empirical knowledge but constitute an interior domain that itself contains a vast store of other types of structures, patterns, knowledge, values, and contents—ranging from logic to mathematics to ethics to linguistics. Empirical-sensory science cannot investigate these domains with its exterior tools; but it would indeed take a village idiot to deny their existence or to deny that other modes of investigation might give access to these extraordinary domains.

AN OPENING TO THE DEEP

Objection number 1—the belief that interior domains have no irreducible reality of their own, and that only sensory-empirical objects are fundamentally real—is actually a notion that few philosophers of science, and few scientists themselves, really believe. Any scientist who uses mathematics already knows that reality is not just sensory. The vast majority of scientists already reject the myth of the given, and all that is required is to keep pointing it out.

The myth of the given is really the myth of exteriors untouched by interiors, of mere objects untouched by subjective and intersubjective structures. It is the myth that there are less than four quadrants to the Kosmos, the myth of the Big One instead of the Big Three. It is the myth at the very core of classical empiricism, positivism, behaviorism, collapsed modernity, and scientism. It is the myth of objects without subjects, of surfaces without depth, of quantity without quality, of veneers without value—the utterly rancid myth that *the Right-Hand world alone is real*. But it is indeed a myth, and the myth is decidedly dead.

Once the myth of the given is exploded, the first major objection of empirical science to interior knowledge is likewise exploded. Science cannot reject a mode of knowing merely because it is interior. That being so, science is then forced to get, shall we say, picky. It must attempt to reject *some* interior modes—such as the contemplative and spiritual—and it can do so *only* by resorting to objection number 2, namely, that these other interior modes have no valid means of verification.

Well, let us see.

11

WHAT IS SCIENCE?

We have seen that the philosophers of science are in wide-spread agreement that empirical science depends for its operation upon subjective and intersubjective structures that allow objective knowledge to emerge and stabilize in the first place. Put bluntly, knowledge of sensory exteriors depends upon nonsensory interiors, interiors that are just as real and just as important as the exteriors themselves. You don't get a message on the telephone, claim the message is real but the phone is illusory. To discredit one is to discredit the other.

If sensory-oriented science is not equipped to investigate these interior domains, it nonetheless cannot deny their existence without denying its own operations. It can no longer claim that the exteriors alone are real. And that, very simply, completely undercuts what we called objection number 1 (the belief that interior domains have no reality of their own). Precisely because empirical science is forced to acknowledge interiors, it cannot dismiss Spirit merely on the basis that Spirit is interior. The first major objection falls.

Thus, if science wishes to continue to deny Spirit, it is forced to retreat to objection number 2 and attempt to deny, not all interiors, but only certain types of interiors, because, it is claimed, these other and "disreputable" interiors—such as spiritual experience—cannot be verified. They are at best merely private modes of knowing; at worst, hallucinations.

Traditionally, what has spooked empirical and positivistic science about these "interiors" is that they cannot be objectified and thus nailed with a sensorimotor hammer, whether that hammer be a telescope, microscope, photographic plate, or whatnot. Thus traditionally, empirical science tended to a simple confusion: it claimed that its basic *methodology* covered all of the real *dimensions* of existence, whereas these are two entirely separate considerations. Once we tease apart the scientific *method* from its application to a particular *domain*, we might find that a certain spirit of scientific inquiry, honesty, and fallibilism can indeed be carried into the interior domains (which science *already* does with its own mathematics and logic). We might find that "science" in the broadest sense does not have to be confined to sensory patches, but might include a science of sensory experience, a science of mental experience, and a science of spiritual experience.

If this were so, it might go a long way to help "unspook" the interiors and set them on a much more reassuring epistemological footing. To do this, we must look a little more closely at what we mean by "science."

THE SCIENTIFIC METHOD

The notion that there exists a single, straightforward "scientific method" has long been discredited. It is almost unanimously acknowledged that there is no algorithm (no set method) for generating theories from data; the very notion was part of the myth of the given. Nonetheless, most philosophers—and certainly most working scientists—have a clear enough idea of what "doing science" actually means; enough, anyway, to differentiate scientific knowledge from poetry, faith, dogma, superstition, and nonverifiable proclamations. The scientific method might be slippery, but it still manages to get a lot of work done—it can, after all, plop a person on the moon, which presumably it could not do if it had no method at all. I believe we can in fact state some of the general ingredients in the scientific method, and I will attempt do so in a moment.

But one of the most interesting things about the scientific method is that nothing in it says that it must be applied *only* to sensory domains or to sensory experience alone. After all, we tend to think of vector analysis, logic, tensor calculus, imaginary numbers, Boolean algebra, and so on as being "scientific" in the broad sense, yet none of those are primarily empirical-sensory. Clearly, "sensory" and "scientific" are not the same thing at all.

Thus, when we look for the defining characteristics of the scientific method, we cannot make "sensory empiricism" one of them. The defining patterns of scientific knowledge must be able to embrace both biology and mathematics, both geology and anthropology, both physics and logic—some of which are sensory-empirical, some of which are not.

Part of the confusion in this area comes from the fact that, historically, "empirical" has been given two broad but quite different meanings. And, I believe, it is an understanding of these two types of empiricism that holds the key to the scientific method.

TWO TYPES OF EMPIRICISM

On the one hand, "empirical" has meant *experiential* in the broadest sense. To say that we have empirical verification simply means that we have some sort of direct experiential evidence, data, or confirmation. To be an "empiricist" in this broad sense simply means to demand *evidence* for assertions, and not merely to rely on dogma, faith, or nonverifiable conjectures.

I have a great deal of sympathy for that position. In fact, using "empirical" in the broad sense of "demand for experiential evidence," I count myself a staunch empiricist. For the fact is, there is sensory experience, mental experience, and spiritual experience—and empiricism in the very broadest sense means that we always resort to *experience* to ground our assertions about any of those domains (sensory, mental, spiritual).

Thus, there is *sensory empiricism* (of the sensorimotor

world), *mental empiricism* (including logic, mathematics, semiotics, phenomenology, and hermeneutics), and *spiritual empiricism* (experiential mysticism, spiritual experiences).

In other words, there is evidence seen by the *eye of flesh* (e.g., intrinsic features of the sensorimotor world), evidence seen by the *eye of mind* (e.g., mathematics and logic and symbolic interpretations), and evidence seen by the *eye of contemplation* (e.g., satori, *nirvikalpa samadhi*, gnosis).

As we will see, the experiential evidence in each of these modes is actually quite *public* or shared, because each of them can be trained or educated with the help of a teacher, and an educated eye is a shared eye (or else it could not be educated in the first place). In all of these ways and more, empiricism in the broadest sense is the surest way to anchor the objective component of truth and the demand for evidence (whether of the exteriors or interiors or both), and thus *empiricism in the broadest sense* will be a crucial aspect of our validity procedures for any domain.

On the other hand, empiricism has also historically been given an extremely narrow meaning, not of experience in general, but of sensory experience alone. Moving from the profoundly important notion that all knowledge must be ultimately grounded in experience, many classical empiricists collapsed this to the absurd notion that all knowledge must be reduced to, and derived from, colored patches. The myth of the given, the brain-dead flatland stare, the monological gaze, the modern nightmare: with this impoverished empiricism, we can have little sympathy.

This dual meaning of "empiricism"—very broad and very narrow—is actually reflected in the extensive confusion about the scientific method itself, and whether it must be "empirical" or not. For the enduring strength of science—the reason it can indeed plop a person on the moon—is that it always attempts, as best it can, to rest its assertions on *evidence* and *experience*. But sensory experience is only one of several different but equally legitimate types of experience, which is precisely why mathematics—seen only inwardly, with the mind's eye—is still considered scientific (in fact, is usually considered extremely scientific!).

When we "do mathematics," we *inwardly perceive*, with the mind's eye, a whole series of symbolic and imaginative events. These are not "mere abstractions"; as any mathematician will attest, they are part of an incredibly rich stream of often quite beautiful images, patterns, scenes, and interior landscapes, which follow what seem at times to be almost divine patterns, exquisitely unfolding before the mind's eye. More astonishingly, many of the patterns in the exterior and sensorimotor world—from the motion of the planets to the speed of falling objects—happen to follow quite precisely these interior mathematical patterns. These are no mere abstractions but profound patterns embedded in the Kosmos itself, yet seen only with the mind's inward eye!

This interior mathematical *experience* is part of the essential ground of mathematical knowledge. We run the equations "through our head" and see if they *make sense*—not sensory sense but mental sense, logical sense (following any number of logics, from Boolean to *n*-dimensional, none of which can be seen with the eye of flesh). In mathematical proofs, we follow a *mental empiricism*, a mental experience, a mental phenomenology, and we see if the patterns connect correctly. We then *check our interior experience with others* who have run the same interior experiment, in order to see if they experienced the same result. If the majority of people who are qualified report the same interior experience, we generally call this a "mathematical proof," and we consider it a case of genuine knowledge.

Thus, a direct, interior, mental *experience* (or empiricism in the broad sense) has guided our every move through the mathematical domain, and these inwardly experiential moves can be checked—confirmed or rejected—by those who have *performed the same interior experiment* (run the proof through their minds).

So the confusion about whether "the scientific method" must be empirical depends entirely on what we mean by "empirical." Do we mean in the broad sense (experience in general), or the narrow sense (sensory experience only, the eye of flesh alone)? My point is that *science cannot mean empiricism in the narrow sense,* because that would rule out mathematics, logic, and most of the conceptual tools of science itself (not to mention psychology, history, anthropology, and sociology).

With all that in mind, let us see if we can abstract the essentials of the scientific method *in the broad sense*, which would be based on *empiricism in the broad sense*. If we can do this—and then further show that this scientific method in the broad sense is applicable to the interior domains in general (as it already is in the case of mathematics and logic)— we will have gone quite a distance in legitimating the interiors themselves (and defusing objection number 2).

We would indeed then have a science of sensory experience, a science of mental experience, and a science of spiritual experience—a monological science, a dialogical science, and a translogical science—a science of the eye of flesh, a science of the eye of mind, and a science of the eye of contemplation—with the traditional concerns of religion joining hands with the assurances of modern science.

THE THREE STRANDS OF VALID KNOWLEDGE

We begin with what appear to be some of the essentials of the scientific method in general. Having extracted these ingredients, the hope is that we will find them equally applicable to the interior domains, thus giving us a methodology that could legitimate the interiors with as much confidence as the exteriors. And the further hope is that, hidden somewhere in those newly legitimated interiors, awaits the awareness of a radiant God.

Here are what I believe are three of the essential aspects of scientific inquiry—what I will also call the "three strands of all valid knowing":

   1. *Instrumental injunction.* This is an actual practice, an exemplar, a paradigm, an experiment, an ordinance. It is always of the form "If you want to know this, do this."

2. *Direct apprehension.* This is an immediate experience of the domain brought forth by the injunction; that is, a direct experience or apprehension of data (even if the data is mediated, at the moment of experience it is immediately apprehended). William James pointed out that one of the meanings of "data" is direct and immediate experience, and science anchors all of its concrete assertions in such data.

3. *Communal confirmation (or rejection).* This is a checking of the results—the data, the evidence—with others who have adequately completed the injunctive and apprehensive strands.

To take them one a time:

In order to see the moons of Jupiter, you need a telescope. In order to understand *Hamlet*, you need to learn to read. In order to see the truth of the Pythagorean theorem, you must learn geometry. If you want to know if a cell has a nucleus, you must learn to take histological sections, learn to stain cells, learn to use a microscope, and then look. In other words, all of those forms of knowing have, as one of their significant components, an *injunction:* If you want to *know* this, you must *do* this.

This is obviously true in the sensory sciences, such as biology, but it is true as well in the mental sciences, such as mathematics. As G. Spencer Brown, in his famous *Laws of Form*, pointed out: "The primary form of mathematical communication is not description, but injunction. In this respect it is comparable with practical art forms like cookery, in which the taste of a cake, although literally indescribable, can be conveyed to a reader in the form of a set of injunctions called a recipe. . . . Even natural science [sensory-empirical] appears to be dependent on injunctions. The professional initiation of the man of science consists not so much in reading the proper textbooks [although that is also an injunction], as in obeying injunctions such as 'look down that microscope' [strand 1]. But it is not out of order for men of science, having looked down the microscope [and apprehended the data, strand 2], now to describe to each other, and to discuss among themselves, what they have seen [strand 3]. Similarly, it is not out of order for mathematicians, each having obeyed a set of injunctions [e.g., imagine two parallel lines meeting at infinity; picture the cross-section of a trapezoid; take the square of the hypotenuse], to describe to each other, and to discuss among themselves, what they have seen [with the eye of mind], and to write textbooks describing it. But in each case, the description is dependent upon, and secondary to, *the set of injunctions having been obeyed first. . . .*" (my italics).

The injunctive strand of knowledge leads to an *experience, apprehension,* or *illumination,* a direct disclosing of the data or referents in the worldspace brought forth by the injunction. Thus, if you want to know if it is raining outside, go to the window and look (the injunction). With this looking or experiencing, there is a direct apprehension ("I see the rain"). This is the immediate data, the direct experience, the intuitive or nonmediated grasp of the moment's appearance. It does not matter in the least if the *immediate* data themselves are actually embedded in chains of *mediated* events (such as culturally molded contexts), because at the moment of apprehension, even mediated events are immediately experienced (or else there would be no experience whatsoever, just endless mediation).

Thus an injunction brings forth or discloses an illumination, experience, or data, and these data are a crucial anchor of genuine knowledge. This also implies that if other competent individuals faithfully repeat the injunction or the experiment ("Go to the window and look"), they will experience roughly the same thing, the same data ("Yes, it is raining outside"). In other words, the illumination or apprehension is then *checked* (confirmed or refuted) by all those who have adequately performed the injunction and thus disclosed the data.

Science, of course, usually includes the formation of hypotheses and the testing of these hypotheses against further data accumulation, but each of those steps also follows the same three strands. The hypothesis is a mental experience that is used to represent various intrinsic features of sensory experience, and both of those—the mental map and the sensory territory—are checked for validity by following the three strands applied to their own domain. The map is thus checked against other maps for coherence, and against other sensory data for correspondence. Each of these checking procedures follows the three strands.

EVIDENCE, KUHN, AND POPPER

Those three strands, I believe, are the essential ingredients of the scientific method (and all valid modes of knowing in general, as I will try to show). This conclusion is bolstered by the fact that, of the three major schools of the philosophy of science that are most influential today—namely, empiricism, Thomas Kuhn, and Sir Karl Popper—this approach explicitly incorporates the essentials of each of them. To take them in that order.

The strength of empiricism is its demand that all knowledge be grounded in experiential evidence, and I agree entirely with that demand. But, as we saw, not only is there sensory experience, there is mental experience and spiritual experience (direct data or direct experience delivered by the eye of flesh, the eye of mind, and the eye of contemplation). Thus, if we use "experience" in its proper sense as direct apprehension, we can firmly honor the empiricists' demand that *all genuine knowledge be grounded in experience*, in data, in evidence. The empiricists, in other words, are highlighting the importance of the apprehensive or *illuminative strand* in all valid knowledge.

But evidence and data are not simply lying around waiting to be perceived by all and sundry, which is where Thomas Kuhn enters the picture.

Kuhn, as we saw, pointed out that normal science proceeds most fundamentally by way of *paradigms* or *exemplars*. A paradigm is not merely a concept, it is an actual practice, *an injunction*, a technique taken as an exemplar for generating data. And Kuhn's point is that genuine scientific knowledge is grounded in paradigms, exemplars, or injunctions. New injunctions disclose new data (new experiences), and this is why Kuhn maintained *both* that science is progressive and

cumulative, and that it shows certain breaks or discontinu-
ities (new injunctions bring forth new data). Kuhn, in other
words, is highlighting the importance of the *injunctive strand*
in the knowledge quest, namely, that data are not simply
lying around waiting for anybody to see, but rather are
brought forth by valid injunctions.

The knowledge brought forth by valid injunctions is indeed genuine knowledge precisely because, contrary to extreme postmodernism, paradigms disclose data, they do not
merely invent it. (The data itself may have been given or
constructed, but the disclosure itself is not merely a construction.) The validity of these data is demonstrated by the
fact that *bad data can indeed be rebuffed*, which is where Popper enters the picture.

Sir Karl Popper's approach emphasizes the importance of
falsifiability: genuine knowledge must be open to disproof, or
else it is simply dogma in disguise. Popper, in other words, is
highlighting the importance of the *confirmation/rejection
strand* in all valid knowledge; and, as we will see, this falsifiability principle is operative *in every domain*, sensory to mental to spiritual.

Thus this overall approach acknowledges and incorporates
the moments of truth in each of those important contributions to the quest for knowledge (evidence, Kuhn, and Popper), *but without the need to reduce those truths to sensory
patches*. The mistake of the narrow empiricists is their failure
to see that, in addition to sensory experience, there is mental
and spiritual experience. The mistake of the Kuhnians is their
failure to see that injunctions apply to all forms of valid
knowledge, not just sensorimotor science. And the mistake of
the Popperians is their attempt to restrict falsifiability to sensory data and thus make "falsifiable by sensory data" the criterion for mental and spiritual knowledge—thus implicitly
and illegitimately rejecting those modes right at the start—
whereas bad data in those domains *are indeed falsifiable*, but
only by further data *in those domains*, not by data from lower
domains!

For example, a bad interpretation of *Hamlet* is falsifiable, not by sensory data, but by further mental data, fur-

ther interpretations—not monological data but dialogical data—generated in a community of interpreters. *Hamlet* is not about the search for a sunken treasure buried in the Pacific. That is a bad interpretation, a false interpretation, and this *falsifiability* can easily be demonstrated by any community of researchers who have adequately completed the first two strands (read the play and apprehend its various meanings).

As it is now, the Popperian falsifiability principle has one widespread and altogether perverted use: it is implicitly restricted *only to sensory data*, which, in an incredibly hidden and sneaky fashion, *automatically bars all mental and spiritual experience from the status of genuine knowledge*. This unwarranted restriction of the falsifiability principle claims to separate genuine knowledge from the dogmatic, but all it actually accomplishes, in this shrunken form, is a silent but vicious reductionism.

On the other hand, when we free the falsifiability principle from its restriction to sensory data, and set it free to police the domains of mental and spiritual data as well, it becomes an important aspect of the knowledge quest in all domains, sensory to mental to spiritual. And in each of those domains, it does indeed help us to separate the true from the false, the demonstrable from the dogmatic.

These three strands, then, will be our guide through the delicate world of the deep interiors, the within of the Kosmos, the data of the Divine, where they will help us, as they do with the exteriors, to separate the dependable from the bogus.

TO GIVE A LITTLE

If science and religion are to be integrated, each must give at least a little, without, however, deforming themselves beyond recognition. We have asked science to do nothing more than expand from narrow empiricism (sensory experience only) to broad empiricism (direct experience in general), which it already does anyway with its own conceptual operations, from logic to mathematics.

But religion, too, must give a little. And in this case, religion must open its truth claims to direct verification—or rejection—by experiential evidence. Religion, like science, will have to engage the three strands of all valid knowledge and anchor its claims in direct experience.

In this chapter, we have looked at "real science." In the next, we will look at "real religion." And perhaps we will find that, just as science can, by its own admission, expand its scope from a narrow empiricism to a broad empiricism, so religion can, as it were, restrict its scope from dogmatic proclamations to direct spiritual experience. In this move, with both parties surrendering an aspect of their traditional baggage that in fact serves neither of them well, science and religion would fast be approaching a common grounding in experiential data that finds the existence of rocks, mathematics, and Spirit equally demonstrable.

12

WHAT IS RELIGION?

"Religion," of course, has many meanings, definitions, and proposed functions. The term has been applied to everything from dogmatic beliefs to mystical experience, from mythology to fundamentalism, from firmly held ideals to passionate faith. Moreover, scholars tend to separate the content of religion (such as belief in angels) from the function of religion (such as maintaining social cohesion), arriving at the awkward conclusion that even when the content may be dubious, the function may be beneficial. We will examine many of these definitions and proposed functions of religion as we proceed.

In the meantime, it will escape no one's attention that in referring to authentic spirituality I am largely excluding the mythological and mythopoetic themes—such as the virgin birth, the bodily ascension, the parting of the Red Sea, the birth of Lao Tzu as a nine-hundred-year-old man, the earth resting on a divine Hindu serpent, the goddess as mythic Gaia—that have formed the substance of the vast majority of the world's religious systems, whether existing in the premodern world or carrying over into the modern.

I am not claiming that these beliefs are unimportant or that they serve no function at all. In fact, they serve a very important developmental or evolutionary function; but as we will see, with the irreversible differentiations of modernity, most of those premodern beliefs and functions of religion are no longer legitimate and can no longer be sustained in modern consciousness (except among those who remain at a premodern level in their own development).

I am therefore claiming that when it comes to a modern science of spirituality (a science of direct spiritual experience and data), those mythological themes—and mythology itself—will form no essential part of authentic spirituality. Isn't this rather drastic? And how many religions would agree with this claim?

As we began to see in the previous chapter, if science and religion are to be integrated, each of them will obviously have to give a little—and yet, I maintain, not so much as to deform them and make them unrecognizable to themselves. We saw that science needs to recognize that its own method rests, not on empiricism in the narrow sense (sensory experience), but on empiricism in the broad sense (experience in general), and this is not a very difficult stretch because virtually all of science's own conceptual apparatus (from logic to mathematics) is *already* empirical in the broad sense.

We simply ask science *to form a more accurate self-image:* to surrender its narrower and inaccurate conception of itself for a broader and more accurate (and already implicitly accepted) conception of itself. Most philosophers of science have already done so, as we saw: "Classical empiricism is a myth."

Likewise, we must ask religion to accept a more authentic self-image of its own possibilities. Particularly in the wake of the irreversible differentiations of modernity, religion must seriously ask itself: What is the actual cognitive content and validity of its claims? Did Moses really part the Red Sea? Was Jesus actually borne by a virgin? Does the earth really rest on a divine serpent? Did creation really occur in six days? Was Lao Tzu actually nine hundred years old when he was born?

If those proclamations are quietly put aside, what is left of religion that it can call its own? And would it still indeed recognize itself?

MYTHOLOGY AND POWER

Religious *mythological* proclamations are clearly *dogmatic*, which means that when they are taken to be literal truths, they are simply asserted without any supporting evidence. As such, they fail the test of the three strands of all valid knowledge. At one time, those beliefs performed various important cultural functions, such as maintaining social cohesion, because they formed the basis of a legitimate (or consensual) intersubjective worldview. But with the differentiations and increased depth of the dignity of modernity, a more sophisticated truth disclosure placed these mythological claims in irreversible doubt.

With each developmental unfolding, the truths of the higher domain place the truths of the lower domains into a profoundly different context, a context that, because it transcends and includes its juniors, also *preserves* and *negates* various features of its predecessors. Modernity *preserved* many of the aspirations, ideals, and values expressed in the best of mythology (such as retribution and justice) but *negated* most of its literal contents (such as the notion that we all actually descended from Adam and Eve).

This is why I depart from most modern sociologists, who generally maintain that mythology has no cognitive value at all (its claims are bogus), yet it nonetheless forms an indispensable social glue and cohesive force for many cultures. This is an incoherent position. Humans cannot live on cognitive falsity alone. Mythology is true enough in its own worldspace; it's just that perspectival reason is "more true": more developed, more differentiated-and-integrated, and more sophisticated in its capacity to disclose verifiable knowledge.

Thus the higher truths of rationality pass judgment on the lower truths of mythology, and for the most part mythology simply does not survive those more sophisticated tests. Moses did not part the Red Sea, and Jesus was not borne by a biological virgin. Those claims, in the light of a higher reason, are indeed bogus.

Of course, premodern revivalists often read deeply metaphorical meanings into mythology (e.g., the virgin birth is really a metaphor for the pure and "immaculate" nature of our higher Self), claiming that it communicates and delivers truths that are higher than reason. This is a double duplicity: first, because this approach actually uses reason to explain some deep truth for which mythology is supposed to be superior, and second, because it then reads this truth into mythology in a way that believers in the myth do not accept at all. For a true believer, the virgin birth is absolutely *not* a metaphor but a concrete, literal, historical fact (one the premodern revivalist actually denies!). The premodern revivalist simply uses the higher powers of reason to read deeper truths into a mythic symbol that itself rarely if ever carried any such meaning for its actual believers, and thus the premodern revivalist, attempting to elevate myth above reason, conceals two deceptions in every utterance: reason is robbed of its actual contribution while mythology is given credit for what it does not possess. This double lie is offered to humanity as a source of spiritual transformation.

But the fact remains: the concrete-literal forms of mythology cannot withstand—and have not withstood—the tests of modernity: those concrete claims are indeed bogus. And *if religion is to survive in a viable form in the modern world, it must be willing to jettison its bogus claims,* just as narrow science must be willing to jettison its reductionistic imperialism.

The real problem is that the mythic, mythological, and mythopoetic approaches to spirituality all involve various types of *mental* forms attempting to explain *transmental* and *spiritual* domains, and however phase-specifically appropriate those approaches might have been for the premodern mythic era, they will no longer work on a collective or even individual level. Mythology will not stand up to the irreversible differentiations of modernity; it confuses prerational with transrational; it fosters regressive ethical and cognitive modes; it hides from any sort of validity claims and actual evidence; and thus avoiding truth, is left only with power as one of its prime motives.

Because evidence undoes mythology, mythology intrinsically hides from evidence. Thus mythology is—and historically has been—a massive source of personal and social oppression. This is why the Enlightenment, as Habermas points out, always understood itself as a *counterforce to mythology*. The clarion call of the Enlightenment was for *evidence*, not for myths, because these myths, despite the lovely halo given them by today's premodern revivalists, were in fact a source of brutal social hierarchies, gender oppression, wholesale slavery, and barbaric torture. "Remember the cruelties!" was indeed the battle cry of the Enlightenment, and for precisely that reason.

The Enlightenment thus maintained that those who come to you with mythology come with hidden (or not so hidden) power drives. Those who attempt to wield these forms of myth hide from evidence, understandably: *to expose their claims to evidence would rob those claims of their power*—and thus rob the owners of those claims of their power, too. Thus hidden from truth and seated in power, they seek to enclose others in that same darkness, usually in the name of their God or Goddess. It is no accident that wars fought in whole or part in the name of a particular mythic Deity have historically killed more human beings than any other intentional force on the planet. The Enlightenment pointed out—quite rightly—that religious claims hiding from evidence are not the voice of God or Goddess, but merely the voice of men or women, who usually come with big guns and bigger egos. Power, not truth, drives claims that hide from evidence.

THE CONTEMPLATIVE CORE

Authentic spirituality, then, can no longer be mythic, imaginal, mythological, or mythopoetic: it must be based on falsifiable evidence. In other words, it must be, at its core, a series of direct mystical, transcendental, meditative, contemplative, or yogic experiences—*not sensory* and *not mental*, but transsensual, transmental, transpersonal, transcendental consciousness—data seen not merely with the eye of flesh or with the eye of mind, but with the eye of contemplation.

Authentic spirituality, in short, must be based on direct

spiritual experience, and this must be rigorously subjected to the three strands of all valid knowledge: injunction, apprehension, and confirmation/rejection—or exemplar, data, and falsifiability.

With the differentiations of modernity, premodern religions of every variety faced an unprecedented situation: precisely because modernity differentiated the value spheres and let them proceed unencumbered and with their own dignity, these newly liberated spheres quickly outpaced in most ways anything the premodern religions could offer. When it came to the world of sensory facts, the answers given by premodern religion (e.g., the earth was created in six days) now faced modern empirical science, and it was simply no contest. When it came to the mental sphere and its operations, religion faced modern developments in mathematics, logic, critical philosophy, philology, and hermeneutics (including the *real* sources of the biblical narratives), and once again premodern religion was no match for the differentiations of modernity.

It is only when religion emphasizes its heart and soul and essence—namely, direct mystical experience and transcendental consciousness, which is disclosed not by the eye of flesh (give that to science) nor by the eye of mind (give that to philosophy) but rather by the eye of contemplation—that religion can both stand up to modernity and offer something for which modernity has desperate need: a genuine, verifiable, repeatable injunction to bring forth the spiritual domain.

Religion in the modern and postmodern world will rest on its unique strength—namely, contemplation—or it will serve merely to support a premodern, predifferentiated level of development in its own adherents: not an engine of growth and transformation, but a regressive, antiliberal, reactionary force of lesser engagements.

But the thorny question remains: Can religion recognize itself if it brackets (or temporarily sets aside) its mythic baggage? For an answer, I suggest we look at the example, not of the followers, but the founders, of the major religions themselves.

REAL AND BOGUS

The first thing we can't help but notice is that the founders of the great traditions, almost without exception, underwent a series of profound *spiritual experiences*. Their revelations, *their direct spiritual experiences*, were *not* mythological proclamations about the parting of the Red Sea or about how to make the beans grow, but rather direct apprehensions of the Divine (Spirit, Emptiness, Deity, the Absolute). At their peak, these apprehensions were about the direct union or even identity of the individual and Spirit, a union that is not to be thought as a mental belief but lived as a direct experience, the very *summum bonum* of existence, the *direct realization of which* confers a great liberation, rebirth, metanoia, or enlightenment on the soul fortunate enough to be immersed in that extraordinary union, a union that is the ground, the goal, the source, and the salvation of the entire world.

And what each of those spiritual pioneers gave to their disciples was *not* a series of mythological or dogmatic beliefs but a series of practices, injunctions, or exemplars: "Do this in remembrance of me." The "do this"—the injunctions—included specific types of contemplative prayer, extensive instructions for yoga, specific meditation practices, and actual interior exemplars: if you want to *know* this Divine union, you must *do* this.

These injunctions reproduced in the disciples the spiritual experiences or the spiritual data of the evolutionary pioneers. In the course of subsequent interior experiments (over the decades and sometimes centuries), these injunctions and data were often refined and sophisticated, with initial or preliminary methods and data polished in the direction of more astute observations. Of numerous examples: the growth and evolution of Hinayana Buddhism into Mahayana Buddhism, which grew and evolved into the magnificent Vajrayana; the exquisite growth of Jewish mysticism through Hasidim and Kabbalah; the great Hindu flowering from the early Vedas to the extraordinary Shankara to the unsurpassed Ramana Maharshi; the six centuries of refinement from Plato to Plotinus.

On the other hand, the moment that any particular spiritual lineage stopped this exploratory and experimental process—that is, the moment it ceased to employ all three strands in the spiritual quest—it began to harden into mere dogma or mythological proclamations, devoid of direct evidence and experience or the power to transform, and it then served merely to translatively console isolated egos in their immortality projects, instead of transcending the ego in the great liberation of a radiant and spiritual splendor.

The conclusion seems obvious: when the eye of contemplation is abandoned, religion is left only with the eye of mind—where it is sliced to shreds by modern philosophy—and the eye of flesh—where it is crucified by modern science. If religion possesses something that is *uniquely* its own, it is contemplation. Moreover, it is the eye of contemplation, adequately employed, that follows all three strands of valid knowing. Thus religion's great, enduring, and unique strength is that, at its core, *it is a science of spiritual experience* (using "science" in the broad sense as direct experience, in any domain, that submits to the three strands of injunction, data, and falsifiability).

Thus, if science can surrender its narrow empiricism for a broader empiricism (which it *already* does anyway), and if religion can surrender its bogus mythic claims in favor of authentic spiritual experience (which its founders uniformly did anyway), then suddenly, very suddenly, science and religion begin to look more like fraternal twins than centuries-old enemies.

For it then becomes perfectly obvious that the real battle is not between science, which is "real," and religion, which is "bogus," but rather between real science and religion, on the one hand, and bogus science and religion, on the other. *Both* real science and real religion follow the three strands of valid knowledge accumulation, while both bogus science (pseudoscience) and bogus religion (mythic and dogmatic) fail that test miserably. Thus, real science and real religion are actually *allied* against the bogus and the dogmatic and the nonverifiable and the nonfalsifiable in their respective spheres.

If we are to effect a genuine integration of science and religion, it will have to be an integration of real science and real religion, not bogus science and bogus religion. And that means each camp must jettison its narrow and/or dogmatic remnants, and thus accept a more accurate self-concept, a more accurate image of its own estate.

THE EYE OF CONTEMPLATION

We have seen that all valid forms of knowledge have an injunction, an illumination, and a confirmation; this is true whether we are looking at the moons of Jupiter, the Pythagorean theorem, the meaning of *Hamlet,* or . . . the existence of Spirit.

And where the moons of Jupiter can be disclosed by the eye of flesh or its extensions (sensory data), and the Pythagorean theorem can be disclosed by the eye of mind and its inward apprehensions (mental data), the nature of Spirit can be disclosed only by the *eye of contemplation* and its directly disclosed referents: the direct experiences, apprehensions, and data of the spiritual domain.

But in order to gain access to any of these valid modes of knowing, I must be *adequate* to the injunction—I must successfully complete the injunctive strand, I must follow the exemplar. This is true in the physical sciences, the mental sciences, and the spiritual sciences. And where the exemplar in the physical sciences might be a telescope, and in the mental sciences might be linguistic interpretation, in the spiritual sciences the exemplar, the injunction, the paradigm, the practice is: meditation or contemplation. It too has its injunctions, its illuminations, and its confirmations, all of which are repeatable—verifiable or falsifiable—and all of which therefore constitute a perfectly valid mode of knowledge acquisition.

But in all cases, we must engage the injunction. We must take up the exemplary practice, and this is certainly true for the spiritual sciences as well. If we do not take up the injunctive practice, we will not have a genuine paradigm, and therefore we will never see the data of the spiritual domain. We will in effect be no different from the churchmen who refused to follow Galileo's injunction and look through the telescope itself.

Let us examine more closely what this spiritual injunction might mean, and why it might indeed constitute a spiritual science.

TRAINING IN SPIRITUAL SCIENCE

Zen Buddhism has a reputation as a "no-nonsense" school of spiritual discipline. It therefore serves well as a classic example of a science of spiritual experience. The following points can as easily be made with Vedanta, Christian contemplation, meditative Taoism, Neo-Confucianism, or Sufi meditation, to name a few. But Zen's "hardheadedness" might make it easier for scientists who are getting the religion tour for the first time and worrying considerably where Mr. Toad's Wild Ride is leading.

A typical Zen story begins with the student earnestly asking the Master a deeply troubling question, such as what is the meaning of life, why am I here, what or where is Buddha, and so on. The Master in turn might ask a counterquestion, some of which might be very straightforward ("Who is it that wants to know?"), but some of which are famously nonsensical ("What is the sound of one hand clapping?"). In a sense, these are all variations on "Show me your spiritual understanding right now! Show me your Buddha nature, right now!"

The Zen Master, of course, will reject every imaginable intellectual response. A clever student might say, "We are all strands in the great Web of Life." That is exactly a wrong answer, because it is a *mental* response, not a directly transmental, transconceptual, or spiritual response. A more advanced student might yawn, jump up and down, or slap the floor. This is at least getting closer, in that the action is direct and immediate, not some sort of mental chatter. But the Zen Master, in all cases, wants to see direct evidence of immediate realization apprehended with the eye of contemplation, not some sort of intellectual philosophy seen with the eye of mind. Any intellectual response will be radically rejected, no matter what its content!

Rather, the new student, in order to gain this spiritual knowledge, must take up an *injunction*, a paradigm, an exemplar, a practice, which in this case is *zazen*, sitting meditation. And—to make a very long and complex story brutally short—after an average of five or six years of this exemplary training, the student may begin to have a series of profound illuminations. And it's very hard to believe that over the years hundreds of thousands of students would go through this extended hell in order to be rewarded only with an epileptic fit or a schizophrenic hallucination.

No, this is Ph.D. training in the realm of spiritual data. And once this injunctive training begins to bear fruit, a series of illuminations or apprehensions—commonly called *"kensho"* or *"satori"*—begin to flash forth into direct and immediate awareness, and this data is then checked (confirmed or rejected) by the community of those who have completed the injunctive and the illuminative strands. At this point, the answer to the question "What or where is Buddha nature (or Godhead or Spirit)?" will become extremely clear and straightforward.

When the Zen Master then asks you where Buddha is, you will directly and immediately give the answer, and if it springs from a deep and spontaneous realization, the Master will recognize it immediately. The answer is not coming merely from some colored sensory patches, nor from some mental symbols or myths or rational abstractions, but directly from a contemplative realization that is so utterly simple and obvious that Zen says it is just like having a glass of cold water thrown in your face.

But the point is, the actual *answer* to the question "Does Spirit exist?"—the technically correct and precise answer is: satori. The technically correct answer is: Take up the injunction, perform the experiment, gather the data (the experiences), and check them with a community of the similarly adequate.

We cannot mentally *state* what the answer is other than that, because if we did, we would have merely words without injunctions, and they would indeed be utterly meaning-less. As G. Spencer Brown said, it's very like baking a pie: you follow the recipe (the injunctions), you bake the pie, and then you actually taste it. To the question "What does the pie taste like?," we can only give the recipe to those who inquire and let them bake it and taste it themselves.

Likewise with the existence of Spirit: we *cannot* theoreti-cally or verbally or philosophically or rationally or mentally describe the answer in any other ultimately satisfactory fash-ion except to say: *engage the injunction.* If you want to *know* this, you must *do* this. Any other approach and we would be trying to use the eye of mind to see or state that which can be seen only with the eye of contemplation, and thus we would have nothing but metaphysics in the very worst sense—statements without evidence.

Thus: take up the injunction or paradigm of meditation; practice and polish that cognitive tool until awareness learns to discern the incredibly subtle phenomena of spiritual data; check your observations with others who have done so, much as mathematicians will check their interior proofs with others who have completed the injunctions; and thus con-firm or reject your results. And in the verification of that transcendental data, the existence of Spirit will become radi-cally clear—at least as clear as rocks are to the eye of flesh and geometry is to the eye of mind.

PROOF OF GOD'S EXISTENCE

We have seen that authentic spirituality is not the product of the eye of flesh and its sensory empiricism, nor the eye of mind and its rational empiricism, but only, finally, the eye of contemplation and its spiritual empiricism (religious experience, spiritual illumination, or satori, by whatever name).

In the West, since Kant—and since the differentiations of modernity—religion (and metaphysics in general) has fallen on hard times. I maintain that it has done so precisely be-cause it attempted to do with the eye of mind that which can be done only with the eye of contemplation. Because the mind could not actually deliver the metaphysical goods, and yet kept loudly claiming that it could, somebody was bound to blow the whistle and demand real evidence. Kant made the demand, and metaphysics collapsed—and rightly so, in its typical form.

Neither sensory empiricism, nor pure reason, nor practical reason, nor any combination thereof can see into the realm of Spirit. In the smoking ruins left by Kant, the only possible conclusion is that all future metaphysics and *authentic spirituality* must offer *direct experiential evidence*. And that means, in addition to *sensory experience* and its empiricism (scientific and pragmatic) and *mental experience* and its rationalism (pure and practical), there must be added *spiritual experience* and its mysticism (spiritual practice and its experiential data).

The possibility of the direct apprehension of sensory experience, mental experience, and spiritual experience radically defuses the Kantian objections and sets the knowledge quest firmly on the road of evidence, with each of its truth claims guided by the three strands of all valid knowledge (injunction, apprehension, confirmation; or exemplars, data, falsifiability) *applied at every level* (sensory, mental, spiritual—or across the entire spectrum of consciousness, however many levels we wish to invoke). Guided by the three strands, the truth claims of real science and real religion can indeed be redeemed. They carry cash value. And the cash is experiential evidence, sensory to mental to spiritual.

With this approach, religion regains its proper warrant, which is not sensory or mythic or mental but finally contemplative. The great and secret message of the experimental mystics the world over is that, with the eye of contemplation, Spirit can be seen. With the eye of contemplation, God can be seen. With the eye of contemplation, the great Within radiantly unfolds.

And in all cases, the eye with which you see God is the same eye with which God sees you: the eye of contemplation.

13

THE STUNNING DISPLAY OF SPIRIT

If there is indeed a genuine spiritual science, what does it disclose? What does it tell us? And can it really be verified?

NARROW SCIENCE AND BROAD SCIENCE

We have seen that both "empiricism" and "science" have a narrow and a broad meaning (or shallow and deep, depending on your metaphor). Broad empiricism is experience in general (sensory, mental, spiritual), whereas narrow empiricism is sensory experience only. Science per se, or the scientific method, consists of the three strands of valid knowing (exemplar, experience, falsifiability). *Narrow science* confines its use of the three strands to sensory experience only (it follows narrow empiricism), whereas *broad science* applies the three strands to any and all direct experience, evidence, and data (it follows broad empiricism).

Modern empirical science tended to reject the interiors because they (mistakenly) appeared opaque to the scientific method. But, as we have seen, the interiors themselves are in fact accessible, not to narrow science, but to broad science, because the interiors of the I and WE can be experientially explored, investigated, reported, confirmed or rejected using the three strands of all valid knowledge accumulation—using, that is, broad science or deep science.

Thus, we are aiming for a broad science of all four quadrants, not a narrow science of the Right-Hand quadrants only. We are looking for a deep science that includes not just the exteriors of ITs but the interiors of I and WE. We are looking for a deep science of self and self-expression and aesthetics; of morals and ethics and values and meaning; as well as of objects and ITs and processes and systems.

Thus, the Big Three—art, objective science, and morals—can be brought together under one roof using the core methodology of deep empiricism and deep science (the three strands of all valid knowledge). *The I and the WE are finally put on an equal footing with the IT*, NOT by reducing the I and the WE to ITs (whether interwoven or holistic or "new paradigm" or otherwise), but by seeing that all three, just as they are, can be equally accessed using the same general methodology: the three strands of broad science. Broad science (or deep science or deep empiricism) can in fact guide our search in each domain, without the necessity to deform one domain to make it "compatible" with the others. The three strands of deep science *separate the valid from the bogus in each quadrant* (or simply in each of the Big Three), helping us to separate not only true propositions from false propositions, but also authentic self-expression from lying, beauty from degradation, and moral aspirations from deceit and deception.

This move *simultaneously* gives to empirical science its nonnegotiable demand that the scientific method be employed for truth accumulation, yet also relieves narrow science from its imperialism by pointing out that the scientific method can apply as fully and as fruitfully to broad empiricism as to narrow empiricism. This brings broad science to the interior domains of direct mental and spiritual experience: shallow science opens to deep science.

With this move, science is both *satisfied* that its central method is still the epistemological cornerstone of all inquiry (without which it will accept no proposed integration whatsoever), yet also *limited* in its imperialism by the recognition that its own narrow empiricism of ITs can gracefully exist alongside the broad empiricism of I's and WE's, since all are equally and confidently covered by broad science.

The four quadrants (or simply the Big Three) can thus be genuinely united, joined, and integrated under the auspices of a deep science that is as operative in profound mystical experience as in geology, as applicable to moral aspirations as to biology, as dependable in hermeneutics as in physics. None of these domains need to be reduced to the others, tortured to fit some "new paradigm," or twisted beyond recognition in order to "fit" some integrative scheme. Each domain, just as it is, is allowed its own dignity, its own logic, its own architecture, its own form and structure and content—yet each is joined and united by the thread of direct experience and evidence, a deep empiricism that grounds all knowledge in experience and all claims in verifiability.

A BROAD SCIENCE OF
EACH QUADRANT

This integration promises to be a genuine unity-in-diversity. The domains are importantly different and are allowed to be so, but their access follows a similar pattern of disclosure and verification or rejection—namely, the three strands of deep science. This unity-in-diversity is like one flashlight investigating different caves: the light is the same, but the actual form of the investigation will take on different contours in each cave. The same light will disclose different territories, as it should be.

Thus, when we apply the three strands of deep science to the Upper-Right quadrant, this gives us the sciences of the *exteriors* of *individual* holons: physics, chemistry, geology, biology, neurology, medicine, behaviorism, and so forth. I have listed these in Figure 13-1, along with some recognized pioneers in these fields.

Applying the three strands of deep science to the Lower-Right quadrant gives us the sciences of the *exteriors* of *com-*

INTERIOR	EXTERIOR
• Interpretive	• Monological
• Hermeneutic	• Empirical, positivistic
• Consciousness	• Form

	INTERIOR	EXTERIOR
INDIVIDUAL	Sigmund Freud	B. F. Skinner
	C. G. Jung	John Watson
	Jean Piaget	John Locke
	Sri Aurobindo	Empiricism
	Plotinus	Behaviorism
	Guatama Buddha	Physics, biology, neurology, etc.
COLLECTIVE	Thomas Kuhn	Systems theory
	Wilhelm Dilthey	Talcott Parsons
	Jean Gebser	Auguste Comte
	Max Weber	Karl Marx
	Hans-Georg Gadamer	Gerhard Lenski

FIGURE 13-1: BROAD SCIENCES
OF THE FOUR QUADRANTS

munal holons: ecology, systems theory, exterior holism, sociology, and so on. In humans, these "exterior sociological" approaches have included those of such notables as Auguste
Comte, Karl Marx, Talcott Parsons, and Niklas Luhmann.

A deep science of the Upper-Left quadrant gives us the
terrain, the data, the contours of the *interiors* of *individual*
holons. In the human realm, this includes not only the more
formal structures that are disclosed inwardly to the mind's
eye—such as logic and mathematics—but also the more per-

sonal contours disclosed by introspective psychology and depth psychology. This is likewise the domain of self and self-expression, art and aesthetics, and mental phenomenology in general.

Moreover, as we will see in a moment, with the *higher stages of interior development*, genuinely spiritual or mystical experiences begin to unfold, and these, too, can be investigated and validated with the three strands of deep science applied to the advanced stages of the Upper-Left quadrant (I did not list any of these higher stages in Figure 5-1, because that figure covers only average evolution up to this point; it does not include higher evolution, which will be discussed in a moment). Pioneers in this quadrant have included Sigmund Freud, C. G. Jung, Alfred Adler, Jean Piaget, St. John of the Cross, St. Teresa of Avila, Ralph Waldo Emerson, Plotinus, Shankara, Chih-I, and Gautama Buddha.

A broad science of the Lower-Left quadrant reveals the *interiors* of *communal* holons, the intersubjective signs, values, shared cultural meanings and worldviews of a given culture. Unlike the social sciences of the Lower Right, which tend to focus on exterior systems and the monological data of a society (its birthrates, population size, dietary patterns, forces of techno-economic production, types of monetary exchange, information flowcharts, feedback loops, and so on, all of which can be described in it-language), *cultural studies* focus on the shared meanings and intersubjective values that act as the interior glue for members of the society (all of which are significantly described in we-language, and therefore must be studied as a "participant observer").

Thus, social system sciences ask, "What does it do?" or "How does it work?," whereas interpretive and cultural sciences ask instead, "What does it mean?" They approach a culture not from the outside, in an objectifying and distancing stance, but rather from the inside, from the within, in a stance of mutual understanding and recognition. Both approaches are useful and necessary, the one investigating the communal holon from the outside (Right Hand), the other from the inside (Left Hand). Pioneers in cultural hermeneutics include

Friederich Schleiermacher, Wilhelm Dilthey, Martin Heidegger, Jean Gebser, Hans-Georg Gadamer, Thomas Kuhn, Mary Douglas, Peter Berger, and Charles Taylor.

We can start to see that although the three strands of deep science (or valid knowledge accumulation) guide our research in each of the quadrants—thus giving us a methodological integrity capable of integrating all four quadrants (or simply the Big Three)—nonetheless, because the quadrants each have very different contours and types of data, we find different "types of truth" in each of the quadrants, and these differences need to be acknowledged and honored. They are the "diversity" part of the unity-in-diversity, and this diversity is every bit as important as the unity.

These different types of equally important truths are referred to as *validity claims*. Each time the same broad science is applied to a different quadrant, it generates a different type of truth: objective truth (behavioral), subjective truth (intentional), interobjective truth (social systems), and intersubjective truth (cultural justness). One method, many truths, each therefore equally dependable.

THE SPIRITUAL DOMAINS

But if that is so, where does Spirit fit into this scheme?

As I began to suggest in the last chapter, there *already* exist numerous spiritual disciplines that carefully follow the three strands of valid knowledge accumulation, disciplines that are therefore, in effect, authentic spiritual sciences (not exterior sciences but interior sciences, following not narrow empiricism but deep empiricism). These spiritual sciences include the contemplative and meditative traditions of a collective humanity, East and West, North and South, traditions that have been carefully collecting interior spiritual data for at least three thousand years, and traditions that, in deep structure analysis, show a surprising unanimity as to the basic architecture of the higher or spiritual stages of human development.

Moreover, the modern discipline known as *transpersonal*

psychology and psychiatry has taken, as one of its tasks, the scientific investigation of these higher stages of human and spiritual development, and it, too, has discovered a striking similarity of these higher stages, across individuals and across cultures. (If you are interested in pursuing the details of these discoveries, excellent transpersonal anthologies include *Paths Beyond Ego*, edited by Roger Walsh and Frances Vaughan; *Textbook of Transpersonal Psychiatry and Psychology*, edited by Bruce Scotton, Allan Chinen, and John Battista; *What Really Matters—Searching for Wisdom in America*, by Tony Schwartz.)

What transpersonal psychology has discovered, and what the contemplative traditions themselves disclose, is that beyond the typical rational-egoic stages of development ("formop" and "vision-logic" in Figure 5-1), there appear to be at least *four higher stages of consciousness development.*

These higher stages have been given many different names; I refer to them as the psychic, the subtle, the causal, and the nondual. Each of these stages appears to have a quite different type of direct spiritual experience associated with it: nature mysticism, deity mysticism, formless mysticism, and nondual mysticism. (For a detailed discussion of these findings, see *Transformations of Consciousness* and *A Brief History of Everything.*)

Now those particular names and experiences are not so important. What is important is that the transpersonal domains themselves appear to consist of at least four major stages of spiritual development, with different types of data and experiences disclosed at each of those verifiable stages.

Here is the point: if, with reference to the Upper-Left quadrant, we take the stages of human development as carefully outlined by modern developmental psychology (and summarized in Figure 5-1, sensation to vision-logic), and if we *add* the four higher and transpersonal stages (psychic to nondual), *the result is exactly the traditional Great Chain of Being* (as shown, for example, in Table 2-1).

THE MEETING OF PREMODERN
AND MODERN

This is fascinating. The deep sciences of the Upper-Left quadrant (from modern developmental psychology to the contemplative sciences) converge on the traditional Great Chain of Being, exactly as disclosed in the core of the premodern religions. The Great Chain, in this regard, receives a stunning vindication by deep science.

But look at what this also means: for each and every one of the wisdom traditions, the Great Chain of Being covered the whole of reality. But we have just seen that, in the light of the differentiations of modernity (the differentiation of the Big Three, or the four quadrants in general), *the Great Chain, apart from its lowest level, actually covers only the Upper-Left quadrant.* It is far from the whole of reality; it is, as it were, merely one fourth of reality!

And that is exactly why the traditional Great Chain did not, and could not, survive the differentiations of modernity. Because the Great Chain, in effect, covered only "one fourth" of the overall Kosmos—namely, the interior dimensions of the individual from prepersonal to personal to transpersonal—it had no conception of, and thus no way to answer, the stunning discoveries in the other three quadrants, including the amazing discoveries about the brain and consciousness (Upper Right), or about how cultural worldviews affect individuals' perceptions (Lower Left), or about how the social conditions of a culture shape the values of its people (the Lower Right). All of those quadrants were either lumped together as "matter" (the Upper Right and Lower Right) or ignored altogether (the Lower Left)—*precisely because, in the premodern view, these were still largely undifferentiated.*

Every one of those quadrants thus launched a series of devastating attacks against the Great Chain, and, for the most part, those attacks were altogether correct. The modern and differentiated disciplines of physics, chemistry, biology, linguistics, hermeneutics, systems theory, philology, semiotics, anthropology, and sociology tore into the premodern

and predifferentiated worldview with a vengeance, and neither the Great Chain, nor the spiritual worldviews associated
with it, ever recovered.

Just as egregious, the Great Chain theorists, to the extent
they acknowledged the Right-Hand quadrants, placed all of
them on the lowest rung in the Great Chain, namely, the material level. All of the higher levels (including vital body and
mind) were thus "transcendent" to the material body. But the
differentiations of modernity disclosed that the "material"
domains are not so much the lowest rung on the great hierarchy as they are the exterior forms of *each and every rung* on
the hierarchy.

(You can easily see this in Figure 5-1. The Right-Hand or
material components are not the lowest level of the hierarchy, they are simply the objective correlates of each and
every Left-Hand component. They are not lower and higher,
but outer and inner. The material neocortex, for example, is
not on the lowest level; it is the correlative, exterior form of
advanced self-reflexive consciousness and is intimately interwoven with it. The Right Hand is not lower than the Left:
they are the exterior and the interior of any given level of existence. This, with few exceptions, the Great Chain theorists
missed entirely, so that the Great Chain remained largely an
Upper-Left quadrant affair, accurate as far as it went.)

Thus, with the differentiations of modernity, we can
clearly see that the traditional Great Chain occupies, as it
were, not all of reality, but one fourth. In other words, *the
Great Chain is now firmly situated within the differentiations of
modernity*, something that had never happened in any premodern culture. The Great Chain is vindicated in its essential
contours, but it is basically situated in the Upper-Left quadrant, taking its place alongside the other three—and equally
important—quadrants, each of which brings its own irreducible truths to bear on the overall picture.

But this is exactly what we set out to do in the beginning
of this book, namely, to find some scheme that could accommodate both premodern and modern worldviews, and thus
integrate religion and science. Since the core of premodern
religion was the Great Chain, and since the essence of

modernity was the differentiation of the value spheres (the Big Three or the four quadrants), then *in order to integrate religion and science, we sought to integrate the Great Chain with the four quadrants.* We have just done so.

Should this integration prove to be sound, as I believe it will, the Great Chain of Being can take its rightful place within the differentiations of modernity. The massive amounts of data from the traditional spiritual sciences can then be correlated and integrated with the equally massive amounts of data from the modern objective sciences (such as biology, neurology, and medicine), the cultural sciences (such as hermeneutics, semiotics, and political theory), and the social sciences (such as systems analysis, ecology, and sociology).

Since these quadrants are all interrelated and mutually interdependent, the Great Chain itself could not exist without them. And it is only by acknowledging, honoring, and including all four quadrants that the long-sought integration of premodern religion and modern science might finally become a reality.

We are now ready to explore exactly how this extensive integration might occur.

PART IV

THE PATH AHEAD

14

THE GREAT HOLARCHY IN THE
POSTMODERN WORLD

This is what we have seen. The Great Chain of Being—from gross body to conceptual mind to subtle soul to causal spirit, with each expanding sphere enveloping its juniors—was the essential core of the world's great wisdom traditions. No major culture in history was without grounding in some version of this Great Holarchy.

Until, that is, the rise of the modern West. In the wake of the Enlightenment, the modern West became the first significant culture to radically deny the Great Nest of Being—or, more specifically, to deny all but its lowest sphere, matter. Gone the mind, gone the soul, gone the spirit, and in their place, the unending nightmare of monochrome surfaces, the disqualified universe of flatland holism, the great and utterly meaningless system of dynamically interwoven ITs. Bertrand Russell was unerringly accurate when he reported that "Blind to good and evil, reckless of destruction, omnipotent matter rolls on its relentless way. All these things, if not beyond dispute, are yet so nearly certain that *no philosophy that rejects them can hope to stand*" (my italics).

Matter, energy, and information—whether atomistic or systems-oriented, whether static or dynamic processes, whether classical thermodynamics or order-out-of-chaos complexity theories—*are all ITs*, and this epidemic "it-ism" (with no room for the I or the WE in their own terms) was the final mark of official modernity, the strange and distorted legacy of the Western Enlightenment. Thus, to put it mildly, the attempt to integrate premodern religion and modern science has been more than a little daunting, since scientific materialism seems so utterly uncompromising, and "no philosophy that rejects [it] can possibly hope to stand."

It is therefore routine, in virtually all of today's attempts to integrate science and spirituality, to claim that the rise of modern science contributed directly to, or even caused, the "disenchantment of the world." The common and widespread view is that the modern West with its modern science, more or less in one major step, massively rejected soul and Spirit, God and Goddess, sacred nature and immortal soul—and left us with the modern wasteland.

The claim that modernity itself, in one very bad move, rejected the spiritual, now dominates most discourse in this area. This is a typical claim: "The desacrimentalization or devaluation of nature that was begun by the scientific revolution was completed by what is called 'the enlightenment.' " That claim has likewise been made by deep ecologists, neopagans, ecofeminists, radical feminists, wiccans, neoastrologers, theosophists, retro-Romantics, and virtually all of the "new-paradigm" theorists.

Yet we have seen that modernity's rejection of the spiritual actually occurred in not one but two quite different moves, one of which was very good and one of which was very bad. By teasing apart these two steps, we were able to distinguish between the dignity and the disaster of modernity, and this allowed us the first genuine foothold in the sought-for integration.

For, as we saw, the rise of modernity in the West was marked most essentially by the differentiation of the cultural values spheres (of art, morals, and science), spheres that, in premodern cultures, tended to be fused, undifferentiated, or indissociated, so that violence or oppression in one sphere tended to bleed into the others. But with the rise of modernity and the differentiation of the spheres, you could look through Galileo's telescope without being burned at the stake, and you could paint nature without the image of a patriarchal God if that was your desire.

But within a century or so, this differentiation of the spheres—which was the enduring dignity of modernity—began to drift into a dramatic dissociation—which was the horrifying disaster of modernity. The Big Three (of art, morals, and science) splintered and fragmented, and this epidemic *alienation* began to invade and corrupt every corner of modernity itself.

Most egregious of all, sensory-empirical and systems science, in league with industrialization—*since both of them were aggressive "it" endeavors*—began to assault and dominate the other spheres. The colonialization and commodification of the I and the WE by the rampant IT came to define the disaster of modernity. The interior domains altogether—consciousness, soul, spirit, mind, values, virtue, meaning—were all reduced to frisky dust, to order-out-of-chaos process ITs. And so, in this fractured fairy tale, the modern West became the first major culture in all of history to deny the Great Holarchy of Being—and in its place, omnipotent matter, atomistic or systemic or informational ITs, the reign of the unending surface.

Precisely because we can now see that this historic denial occurred, not in one but in two steps, we can more accurately see what of modernity must be treasured—*and therefore integrated*—and what can be discarded. What needs to be integrated is not the dissociations but the differentiations of modernity, for not only do these define the dignity of modernity, they are an *irreversible* part of the evolutionary process of differentiation-and-integration. Modernity has *already* given us these irreversible differentiations—we couldn't undo them even if we tried. What is now required is their *integration*, or the inclusion of all three value spheres (or all four quadrants) in a more encompassing embrace. It is not necessary to attempt to integrate spirituality with the collapsed Kosmos or with the disaster of modernity; it is necessary only to integrate spirituality with the differentiated Kosmos or the dignity of modernity.

Thus, precisely because the essence of the premodern religions was the Great Chain, and because the essence of modernity was the differentiation of the value spheres (the four quadrants, or simply the Big Three), then in order to integrate modern science and premodern religion it is necessary to *integrate the Great Chain with the four quadrants*.

We did exactly that in the last chapter. Here are some of the implications of what we found.

LEVEL AND DIMENSION

Using the three strands of all valid knowledge (paradigm, experience, falsifiability), we were able to suggest a way to integrate the four quadrants with the traditional Great Holarchy of Being. In doing so, we found that each level in the traditional Great Chain is not a single, uniform, monolithic plane (as was traditionally thought), but rather, *each level of the Great Chain actually consists of at least four dimensions or four quadrants*. Each level has a subjective, objective, intersubjective, and interobjective dimension—intentional (Upper Left), behavioral (Upper Right), cultural (Lower Left), and social (Lower Right).

The words "level" and "dimension" are deliberately chosen. In a five-story building, each of its floors is a level, with some floors being higher or lower than others. But the length and width of each floor are its dimensions; neither length nor width is better than the other, and both dimensions are equally present and equally important—you can't have one without the other.

Thus, if we picture the Great Chain as body, mind, soul, and spirit, then *each of those levels* has an intentional, behavioral, cultural, and social dimension. You can see many of these in Figure 5-1, which covers evolution up to the mental levels, and I will give examples of the higher levels as we proceed.

One of our scales therefore involves vertical levels (the traditional Great Chain), and one involves the horizontal dimensions present on each and every level (the four quadrants). In integrating these two scales, we therefore look for the four quadrants (or simply the Big Three) on each of the levels of the traditional Great Holarchy of Being (but *only* to the extent that those levels themselves pass the test of deep science; any "levels" that do not pass the test of deep science, we are not obliged to integrate, for we then have no guarantees that they are real or genuine, and we are not interested in integrating dogma).

Thus, if we continue to use the simple version of the Great Chain—body, mind, soul, and spirit—and if, also for convenience, we shorten the four quadrants to the Big Three (of art, morals, and objective science), then we would have four levels with three dimensions each: the art, morals, and science of the sensory realm; the art, morals, and science of the mental realm; the art, morals, and science of the soul realm; and the art, morals, and science of the spirit realm.

What remains is to give some concrete examples of each of those twelve domains. I am going to give several specific details, some of which you might agree with, some of which you might not; I would ask that you simply take this as a series of examples of how we might proceed in this multidimensional endeavor. If you wish to use other details, or different ones altogether, fine. Don't let my particular details detract from the general procedure. Also, I am going to make these examples deliberately brief and sketchy, so as to not crowd the reader with my own version of events.

LEVELS OF ART

The four levels we are using, in this simplified account, are the sensorimotor, the mental, the subtle soul, and the causal spirit. Each level, as always, transcends and includes its predecessors, so there is nothing mutually exclusive about any of these levels. It is simply that each senior level possesses emergent qualities not found in its juniors, and the art of each level often takes these new, emergent, and defining characteristics as the topic for aesthetic appreciation, *thus giving each level of art a very distinctive stamp* (precisely the same is true for the levels of morals and of science, as we will see). I will use the visual arts, but any will do.

The art of the sensorimotor world takes as its content or referent the sensory world itself, as perceived with the eye of flesh, from realistic impressions to landscapes to portraiture. This is "objective" art or representational art, and whether the art objects are bowls of fruit, landscapes, industrial towns, nudes, railroad tracks, mountains, or rivers, they are all sensorimotor objects. Typical examples include the realists, the Impressionists, and the entire tradition of naturalism.

The art of the mental domain takes as its referent the actual contents of the psyche itself, as interiorly perceived with the eye of mind. The Surrealists are the most obvious; but conceptual art, abstract art, and abstract expressionism are also typical examples. Marcel Duchamp summarized the general point: "I wanted to get away from the physical aspect of painting. I was much more interested in recreating *ideas* in painting. I wanted to put painting once again at the service of the mind"—and not simply the eye of flesh.

But this is not "mental abstraction" in the dry sense. The inward empiricism of the eye of mind—from mathematics to mental art—is actually experience in some of its deepest, richest, most intense textures. As Constantin Brancusi almost screamed out, "They are imbeciles who call my work abstract; that which they call abstract is the most realist, because what is real is not the exterior form but the idea, the essence of things." Mental art attempts to give visual expression to just those ideas and essences.

The art of the subtle level takes as its content or referent various illuminations, visions, and archetypal forms, as inwardly and directly perceived with the beginning eye of contemplation (or transpersonal awareness by whatever name). It is, we might say, *soul art*, as František Kupka stated, "Yes, [this] painting means clothing the processes of the human soul in plastic forms."

This means, of course, that the artists themselves must have evolved or developed into the subtle domain, as Wassily Kandinsky knew: "Only with higher development does the circle of experience of different beings and objects grow wider. Construction on a purely spiritual basis is a slow business. The artist must train not only his mind *but also his soul*" (my italics).

In the Eastern traditions, one of the main functions of this soul art is to serve as a support for contemplation. In the extraordinary tradition of Tibetan *thangka* painting, for example, the buddhas and bodhisattvas that are depicted are not symbolic or metaphoric or allegorical, but rather direct representations of one's own subtle-level potentials. By visualizing these subtle forms in meditation, one opens oneself to those corresponding potentials in one's own being.

The point is that soul art, of any variety, is not metaphoric or allegorical; *it is a direct depiction of the direct experience of the subtle level.* It is not a painting of sensory objects seen with the eye of flesh, and it is not a painting of conceptual objects seen with the eye of mind; it is a painting of subtle objects seen with the eye of contemplation.

That means that artist and critic and viewer alike must be alive to that higher domain in order to participate in this art. As Brancusi reminded us, "Look at my works until you see them. Those who are closer to God have seen them." As Kandinsky put it, the aim is to "proclaim the reign of Spirit . . . to proclaim light from light, the flowing light of the Godhead," all seen, not with the eye of flesh or the eye of mind, but with the eye of contemplation, and then rendered into artistic material form as a reminder of, and a call to, that extraordinary vision.

As the eye of contemplation deepens, and consciousness evolves from the subtle to the causal (and nondual), subtle forms give way to the formless (e.g., *nirvikalpa, ayin, nirodh*) and eventually to the nondual (*sahaja*), which I will together treat as the domain of pure Spirit. The art of this domain takes no particular referent at all, because it is bound to no realm whatsoever. It might therefore *take its referent from any or all levels*—from the sensorimotor/body level (such as in a Zen landscape) to the subtle and causal levels (such as in Tibetan *thangkas*). What characterizes this art is not its content, but the utter absence of the self-contraction in the artist who paints it, an absence that, in the greatest of this art, can at least temporarily evoke a similar freedom in the viewer (which was Schopenhauer's profound insight about the power of great art: it brings transcendence).

But all we need note is that the aesthetic-expressive dimension—the dimension of subjective intentionality and individual interiors—can express and represent any of the levels of the Great Chain, from gross to subtle to causal to nondual, depending upon which level the artists themselves are alive to.

Art, then, is one of the important dimensions of every level in the Great Holarchy of Being. Art is the Beauty of Spirit as it expresses itself on each and every level of its own manifestation. Art is in the eye of the beholder, in the I of the beholder: Art is the I of Spirit.

LEVELS OF MORALS

Developmental psychologists have charted the major stages of moral development in both men and women, and, although the details vary considerably, there is a widespread and general consensus that moral development moves from stages that are *preconventional* (sensorimotor, hedonistic, egocentric, magic-impulsive) to *conventional* (conformist, sociocentric, mythic-membership) to *postconventional* (world-centric, rational-centauric, universal). You can see many of these on Figure 5-1. Carol Gilligan has suggested that men progress through this hierarchy with an emphasis on justice and rights, whereas women tend to develop through the same hierarchy with an emphasis on relationship and care (Gilligan's three hierarchical female stages she calls selfish, care, and universal care, which are the preconventional, conventional, and postconventional stages).

None of these moral structures can be exteriorly seen, of course, because they are interior structures, but structures that nonetheless govern an individual's behavior in the sensory-empirical world. They are, we might say, as real as logic or linguistics or any other authentic interior domain, and they can be studied (and validated) by *a deep science of the intersubjective world*, which is exactly what Piaget, Kohlberg, and Gilligan did (to name a very few).

Both Kohlberg and Gilligan (and, in fact, a now-extensive

number of major moral theorists) have suggested that there is a still higher stage of moral development, which Kohlberg called "universal spiritual." Transpersonal researchers, based on an increasing body of evidence collected by a deep science of the interior, have further suggested that what Kohlberg called "one" spiritual stage has at least three or four subdivisions (as we saw). Let us, for this brief example, simply say that the evidence at this point strongly suggests that there are at least two higher stages of moral development, which I will again simply call subtle soul (saint or bodhisattva) and causal spirit (sage or *siddha*).

The morals of the subtle-soul level, the bodhisattva level, typically involve the deep aspiration to gain enlightenment for all sentient beings (literally). This extraordinary aspiration, which arises spontaneously from the depths of the soul, is based on the growing perception that all sentient beings are direct manifestations of the Divine, and thus are to be treated as manifestations of one's own deepest Being and Self.

The morals of the causal-spirit level, the sage level, involve the paradoxical aspiration to free all sentient beings by realizing that all beings are always already and eternally free. This direct realization of the radically self-liberated nature of all manifestation is behind some of the most sublime (and paradoxical) of the spiritual sciences, and stands as a self-confirming testament of Spirit's timelessly free nature.

What all of that tells us, of course, is that there are levels of moral development, and these levels appear to span the Great Chain from body (hedonistic, preconventional) to mind (conventional and postconventional) to subtle (saintly) to causal (sagely).

Morality, in other words, is one of the important dimensions of every level in the Great Holarchy of Being. Morals are the intersubjective form of Spirit, the Good of Spirit, as it expresses itself on each and every level of its own manifestation. Morals are the We of Spirit.

THE NEW ROLE OF SCIENCE

By levels of science I mean levels of objective, exterior, sensory-empirical science. We are not at this point talking about broad science or deep science (the three strands of all valid knowledge wherever they are applied, interior or exterior). We are talking about traditional empirical science. For the crucial point is that sensory-empirical science, although it cannot see into the higher and interior domains on their own terms, can nonetheless register their empirical correlates. The whole point about the differentiations of modernity is that all interior events have exterior correlates (all holons have both a Left- and a Right-Hand dimension), and this dramatically but *dramatically* changes the role of sensory-empirical science itself.

For objective empirical science is no longer relegated to the bottom rung of the hierarchy (which the traditional approach gave it and which contemporary epistemological pluralism still gives it); rather, empirical science is accessing the *exterior* modes of *all of the higher levels as well*. This moves empirical science off of the bottom level of the Great Chain and places it on the exterior side of each level of the Great Chain. (You can see this in Figure 5-1.) Thus, objective empirical science does not give the whole story, but neither does it have nothing to say about the higher domains, which is the untenable stance of both traditional and contemporary epistemological pluralism (and the crushing error that contributed to the collapse of the Great Chain).

Moreover, this allows us to thoroughly "ground" or "embody" metaphysical or transcendental claims, in effect providing a seamless union of transcendental and empirical, otherworldly and this-worldly. For the higher levels themselves are not *above* the natural or empirical or objective, they are *within* the natural and empirical and objective. Not on top of, but alongside of. Spirit does not physically rise above nature (or the Right-Hand world); Spirit is the interior of nature, the within of the Kosmos. We do not look up, we look within.

This union of Left and Right, interior and exterior, is a

type of transcendental naturalism or naturalistic transcendentalism—a union of otherworldly and this-worldly, ascending and descending, spiritual and natural—a union that avoids, I believe, the insuperable difficulties of either position taken alone.

We saw that one way to summarize the premodern worldview is that it largely emphasized the interior domains (the Great Chain itself, except for its lowest level, is entirely interior and transcendental), and one way to summarize the modern worldview is that it is largely exterior (naturalistic and Right-Hand-oriented). Thus, a type of transcendental naturalism, uniting Left and Right, interior and exterior, transcendental and empirical, is therefore just another way to summarize the marriage of the best of premodern wisdom and modern knowledge.

LEVELS OF SCIENCE

With that prologue, we can now look at the actual levels of sensory-empirical science. In one sense, of course, there are no levels of sensory-empirical science: it simply registers the facts of the sensorimotor world, period. That is true enough; but these "sensorimotor facts" are the exteriors of interiors that are themselves graded in value and meaning and morals and art, and empirical science is perfectly designed to spot the exterior correlates of these interiors.

Here is a simple example. In 1970, R. K. Wallace published "Physiological Effects of Transcendental Meditation" in the prestigious journal *Science*. Wallace's (and others' subsequent) research demonstrated that people in a meditative state display very real and sometimes very dramatic changes in the body's physiology, including everything from blood chemistry to brain-wave patterns. On the basis of this repeatable data, Wallace concluded that the meditative state is a "fourth state of consciousness," as real as the waking, dreaming, and deep sleep states (because, for example, all four states have signature brain patterns as disclosed on an EEG machine).

This research arguably did more to legitimize the meditative state (at least for the Western mind) than all the Upanishads put together. For this research clearly demonstrated that, whatever else meditation is, it is no mere subjective fantasy, ineffectual daydreaming, or inert trance. It produces dramatic and repeatable changes in the entire organism, and most significantly in the electrical patterns of the brain itself, presumably the seat of consciousness.

The question then arises, "But what is the actual meaning of this fourth state? What does it tell us?" And the only possible answer is "Enter that state yourself and find out." For the almost universal consensus of those who do is that this state begins to disclose the Divine.

"Ah," retorts the empirical scientist, "the EEG did not show that the meditators were seeing Spirit or the Divine or some sort of genuinely mystical state. All the EEG showed was that there are empirical differences in the brain waves of the meditative state. You have no right to conclude that the meditative state is a Divine reality, or a higher reality, or that it is in some way more real than the other states."

Correct, but that statement is true for all states on the EEG. When you are in the dream state, perhaps dreaming of seeing a unicorn, the EEG will register a particular pattern. When you wake up, the EEG will register a different pattern. *Subjectively*, you realize that the unicorn of the dream state does not really exist, and so you say the waking state is real and the dream was not real. *But the EEG registers each of them as equally real.* The *objective* machine cannot decide on *subjective* realities, only on the empirical or exterior correlates of those realities.

In other words, the empirical machine gives us the *quantitative* (Right-Hand) but not the *qualitative* (Left-Hand) aspects of these different states. And nothing on the machine says, or can say, that one state is more real or more valuable or more meaningful than another; it can only say that one is different from another. It cannot say that compassion is better than murder, or that truthfulness is better than deceit, or that tolerance is better than bigotry, or that care is better than neglect, only that they are different. The machine can register only changes in size, magnitude, and quantity—all valueless in themselves—whereas the qualitative differences are seen *only* by the inward eye of mind or the inward eye of contemplation, all of which equally register on the neutrally objective machine.

Thus, the empirical scientist is right that the EEG will not say that this fourth state of consciousness is more real (or is disclosing higher realities) than the other states. But neither will the EEG machine say that waking is more real than dreaming or compassion better than murder. If empirical scientists maintain that waking is more real than dreaming, or compassion better than murder, or tolerance better than bigotry, then they will likewise have to hold open the possibility that the meditative state is an opening to the Divine even more real than waking, because that is exactly what is subjectively announced in all of those cases.

And they can actually *check (or refute) that claim using deep science:* namely, take up the injunction or paradigm of meditation; gather the data, the direct experience, the apprehensions that are disclosed by the injunction; compare and contrast the resultant data with that of others who have completed the first two strands. (Those who refuse this injunction are simply not allowed to vote on the truth of the proposition, just as the churchmen who refused to look through Galileo's telescope were not competent to form an opinion about the existence of the moons of Jupiter.)

Of those who do take up the injunction, the strong consensus is that, in this fourth state of consciousness, qualities and insights and freedoms most often characterized as "spiritual" come increasingly to the fore. An expanded sense of self, consciousness, compassion, love, care, responsibility, and concern, tend gradually but insistently to enter awareness. (These claims, too, can—and have—been subjected to empirical and phenomenological tests. See *The Eye of Spirit* for a summary of this research.)

In short, it appears that the very contours of the Divine begin more clearly and more intensely to manifest through this fourth state of consciousness. Subjectively, it is experienced as an increase in precisely those qualities often termed "spiritual" (from awareness to love to compassion), while objectively, it registers in a distinctive set of physiological changes in the organism, including signature brain-wave patterns.

As for the meditative state itself, recent research has begun to reveal numerous levels or sublevels of "the" meditative state, each of which has a distinctive brain-wave pattern (or other empirical correlates). I will, again, simply use two meditative states, traditionally referred to as *"savikalpa samadhi"* and *"nirvikalpa samadhi."*

Savikalpa is "meditation with form," and *nirvikalpa* is "formless meditation." *Savikalpa* produces *subjectively* various displays of archetypal illumination, expansive states of deeply felt love and compassion, and profound motivations to be of service to others; while *objectively* there tends to be brain-hemispheric synchronization (among other things).

Nirvikalpa produces *subjectively* complete cessation of all mental activity, a radically *formless* consciousness that at the same time is experienced as immense (even infinite) freedom and boundless existence, the great Abyss or Emptiness from which all manifestation emerges; while *objectively* there tend to be several rather striking changes in the empirical organism, one of the most stunning of which includes, on occasion, the almost complete cessation of alpha, beta, and theta brain waves, but a large increase in delta waves (usually associated only with deep, dreamless sleep, except that in this case the subject is wide awake and superalert).

Thus, when we talk about *levels of sensory-empirical science,* we mean levels of the interiors (seen with the eye of mind or the eye of contemplation) as they register in the objective and exterior world (seen with the eye of flesh or its extensions—seen, that is, by modern empirical science). We mean the Right-Hand correlates of the Left-Hand worlds (precisely because all holons, without exception, have both Left- and Right-Hand dimensions).

Objective science, then, is one of the important dimensions of every level in the Great Holarchy of Being. Science is the exterior of Spirit, the objective Truth of Spirit, the surface of Spirit, as it expresses itself on each and every level of its own manifestation. Science is the It of Spirit.

THE FACES OF SPIRIT

We have seen that each vertical level of the Great Holarchy has four horizontal dimensions or quadrants—intentional, behavioral, cultural, and social—or simply the Big Three of art, morals, and science; the Beautiful, the Good, and the True; I, WE, and IT.

The Good, the True, and the Beautiful, then, are simply the faces of Spirit as it shines in this world. Spirit seen subjectively is Beauty, the I of Spirit. Spirit seen intersubjectively is the Good, the We of Spirit. And Spirit seen objectively is the True, the It of Spirit.

From the time before time, from the very beginning, the Good and the True and the Beautiful were Spirit whispering to us from the deepest sources of our own true being, calling to us from the essence of our own estate, a whispering voice that always said, love to infinity and find me there, love to eternity and I will be there, love to the boundless corners of the Kosmos and all will be shown to you.

And whenever we pause, and enter the quiet, and rest in the utter stillness, we can hear that whispering voice calling to us still: never forget the Good, and never forget the True, and never forget the Beautiful, for these are the faces of your own deepest Self, freely shown to you.

THE INTEGRAL AGENDA

If we have seen a way to integrate the Great Holarchy of Being with the differentiations of modernity, thus integrating premodern religion and modern science, the question is, what next?

THE PRENUPTIAL AGREEMENT

We have seen that the three strands of deep science (injunction, apprehension, confirmation; or paradigm, data, falsifiability) apply not only to exterior experience; they are the means whereby we decide if a particular interior experience carries genuine knowledge and cognitive content, or whether it is merely hallucinatory, dogmatic, bogus, idiosyncratic, or personal preference. Any interior experience that passes the test of deep science may be provisionally regarded as *genuine knowledge*—that is, it tells us something *real*, something *actual*, about the contours of the Kosmos.

Although many of the claims of the premodern religions cannot pass the test of deep science—and therefore must, at this point, be considered dogmatic, nonverifiable, or bogus— nonetheless the esoteric core of the premodern religions consists not of a series of mythic and nonfalsifiable beliefs, but a series of contemplative practices, actual interior experiments in consciousness, grounded in direct experience. Yoga, *zazen*, *shikan-taza*, *satsang*, contemplative prayer, *zikr*, *daven*, tai chi—these are not beliefs, these are injunctions, exemplars, practices, paradigms.

Thus, Zen and the great contemplative traditions are, in every sense of the word, a deep science of the spiritual interiors, and they have universally concluded that there are *levels of interior experience*. These levels of consciousness are, of course, the Great Chain of Being. Specific mythic beliefs vary dramatically from religion to religion, and it is virtually impossible to construct a universal theology based on these often wildly differing myths. But the Great Holarchy, in one form or another, is the single common framework found in virtually all of the major premodern religions, and thus it is the Great Chain itself that must be included in the long-sought integration.

It is exactly here that the various religions must carefully focus their concerns and make their own modest compromises. Religions the world over will have to *bracket their mythic beliefs*, beliefs such as Moses parting the Red Sea or Lao Tzu being nine hundred years old at birth. I am not asking the more fundamentally oriented faithful to reject those beliefs, merely to set them aside for a moment. For it is abundantly clear that we cannot have any sort of integration of modern science with those specific mythic beliefs. In fact, we cannot have any sort of common ground among the world's religions themselves based on mythic beliefs. For, as we were saying, the myths differ so much in details and content that if one of the ten thousand myths is right, 9,999 are dead wrong. This is no way to build a consensus.

Instead, each religion needs to focus on those aspects of its tradition that were disclosed by its own deep science of the interiors, whether the contemplative prayer of St. Teresa of Avila, the yoga of Patanjali, the vision quest of tundra shamanism, the *zikr* of Rumi, the self-inquiry of Sri Ramana Maharshi, the *shikan-taza* of Bodhidharma, the contemplation of Isaac of Akko, or the meditation of Lady Tsogyal and Padmasambhava, to name a very few. We have been summarizing all of those interior sciences by saying that they universally point to the Great Nest of Being, body to mind to subtle soul to causal spirit, by whatever names.

Each religion can, with modest discomfort, look to the Great Chain in its own tradition, and temporarily bracket its specific, exclusive, proprietary, dogmatic, mythic beliefs. Those beliefs may be true, they may not be true—but so far, they have dramatically failed to pass the test of deep science. Therefore they are not something that science itself (narrow or deep) will accept as valid knowledge, and thus they are not going to be a part of any marriage that science will accept.

But the deep sciences of the interior domains, disclosed by direct experiential evidence and data, evoked by repeatable injunctions, and open to confirmation or rejection by a community of the adequate—those deep sciences are the core of the great wisdom traditions and the core of the Great Chain, and those deep sciences of the spiritual interiors are precisely the genuine knowledge that religion, holding its head high, can bring to the integrative table.

The Great Holarchy itself is more than enough to get the conversation started, and more than enough to serve as a frame to anchor any ongoing integration. *Coupled with the differentiations of modernity and submitted to the tests of deep science*, the Great Chain and its newfound validity ought to be enough foundation for any religion, and on that frame proponents can hang whatever mythic beliefs they want, as long as they do not expect any form of science or any other religion to acknowledge them.

At the same time, this does not mean that we will lose all religious differences and local color and fall into a uniform mush of homogenized New-Age spirituality. The Great Chain is simply the skeleton of any individual's approach to the Divine, and on that skeleton each individual, and each religion, will bring appropriate flesh and bones and guts and glory. Most religions will continue to offer sacraments, solace, and myths (and other translative or horizontal consolations), in addition to the genuinely transformative practices of vertical contemplation. None of that necessarily needs to change dramatically for any religion, although it will be set in a larger context that no longer demands that its myths be the only myths in the world.

EVOLUTION

Religion will also have to adjust its attitude toward evolution in general. I maintain that, contrary to all appearances, this is a *modest* adjustment, because the Great Chain itself is already fully compatible with an evolutionary view. As has often been pointed out, evolution is actually not much more than the Great Chain *temporalized*. That is, if you look at the traditional Great Holarchy as presented by, say, Plotinus or Aurobindo (Table 2-1), it becomes obvious that the *levels of the Great Chain* are actually some of the *major stages of evolution*. As you can see in Figure 5-1, science tells us that the universe has evolved from matter to sensations (in neuronal organisms) to perceptions (with the emergence of the neural cord) to impulses (in reptiles) to images (in mammals) to concepts (in humans). And those are exactly the beginning levels of the Great Chain itself. As the Idealists pointed out, the Great Chain is not given all at once, it unfolds (or evolves) over time; and the stages of evolution given by science closely match the corresponding stages given by the Great Chain theorists.

Thus, to the extent religions bracket their mythic beliefs and focus on their esoteric core (the Great Chain), an acceptance of evolution is a modest adjustment indeed. In fact, Aurobindo has already brought Vedanta (and the entire sweep of Indian philosophy) into an evolutionary accord. Abraham Isaac Kook has already pointed out that "The theory of evolution accords with the secrets of Kabbalah better than any other theory." The great Idealists have already cleared the way for an evolutionary spirituality. And has not the Pope himself finally declared that "evolution is more than a hypothesis"?

What makes it especially hard for some religions to come to terms with evolution is not only their reliance on dogmatic mythic beliefs, but also their commitment to a retro-Romantic view. This view tends to confuse the differentiations (and dignity) of modernity with the dissociations (and disaster) of modernity, and thus it tends to see in modernity nothing but an antispiritual, antireligious desacralization of the world, with the modern West being akin to the Great Satan.

But, as we saw, the modern West is actually an intense combination of good news, bad news. The self or *subject* of rationality was *deeper* than the subject or self of mythology (i.e., the mental-egoic self has *more depth* than the mythic-membership self, because it transcends and includes the essentials of its predecessor). However—solely because of the collapse of the Kosmos—the *object* of rationality (which was confined to sensorimotor flatland) was much less deep than the object of mythology (which was the Divine order, however crudely or anthropomorphically depicted). Thus, *a much deeper subject confined its attention to a much shallower object*. And there, in a nutshell, the combination of dignity and disaster that is the paradox of modernity: a deeper subject in a shallower world.

But retro-Romanticism of every variety imagines that the mythological era itself contained deeper subjects that were subsequently lost and must be regained (in a "mature" form), and that profound error at the very heart of Romanticism guarantees that it and its allies, failing to see that there is no future in the past, will forever be at odds with the general thrust of evolution itself. To the extent that a religion pledges allegiance to a mythic Eden in any actual sense, it will have insuperable difficulty participating in the integration of modern science and spirituality.

The evolutionary or developmental view does not simply praise one epoch and condemn another. Rather, each epoch, each era, each stage of cultural evolution brings with it important truths, valuable insights, and profound revelations. The general evolutionary or developmental view, precisely because it transcends and includes the important truths of each and every one of its stages, takes all important truths with it, enfolded in its own ongoing embrace, and thus honors and includes more truths than any of the alternatives. This means that an evolutionary view is the most viable chariot for a truly integrative stance, extending an embrace that, by any other name, is genuinely compassionate. And if pathology is to be avoided, these truths must be taken up and included in subsequent stages of evolution. Each stage is true, each succeeding stage is "more true": it contains the previous truths and then adds its own, emergent, novel truths, thus both including and transcending its predecessors.

This is not elitist, and it offers no reason for any epoch (even ours) to picture itself as privileged, because it, too, is destined to pass, to be transcended and included in tomorrow's greater embrace. We are, all of us, tomorrow's food. Thus, not only does a developmental or evolutionary view generously give to each period its own important truths, it gives to the present its own appropriate humility.

DEEP SCIENCE RESEARCH

One of the most pressing agendas for this integral view is what might be called an "all-level, all-quadrant" approach to research. This research would *attempt to investigate the various phenomena in each of the four quadrants*—subjective states, objective behavior, intersubjective structures, and interobjective systems—*and correlate each with the others*, without trying to reduce them to the others. This integral approach is a harmonization of the broad sciences of all of the levels in each of the quadrants: thus, "all-level, all-quadrant." Let me give two or three quick examples, relating specifically to psychological and spiritual growth, to show what might be involved.

I mentioned earlier that evolution in general is not much more than the Great Chain temporalized. If we again take Plotinus and Aurobindo as representative examples (see Table 2-1), we can see that evolution so far has unfolded about the first three fourths of the Great Chain, from matter to life function to sensation to perception to emotions to images to concepts to formop to vision-logic. But what about the highest fourth, the stages beyond reason? What about the subtle and causal, the Overmind and the Supermind? If evolution unfolded the first three fourths, is not there every reason to suppose that it will eventually unfold the highest fourth as well? And that, therefore, the levels of soul and Spirit, Overmind and Supermind, lie not in our collective past but in our collective future? And that true religion, far from being a reactionary force yearning for a lost yesteryear, would become, for the first time in history, the vanguard of a progressive, liberal, and evolutionary force?

That, of course, was one of the basic insights of Idealism, and, as far as it goes, there are good reasons to suppose it might indeed be true. This, no doubt, was also one of the great appeals of Father Pierre Teilhard de Chardin, whose notion of the Omega point (of Christ consciousness) as a future attractor for present evolution—a notion borrowed from Schelling and Hegel—freed many Christians from the impossible mythic belief in a literal Garden of Eden and a morbid fixation (a Romantic death wish) to the long-deceased past. This idea is likewise behind the extraordinary integral yoga of Sri Aurobindo, arguably the greatest spiritual-evolutionary theorist.

Should it prove to be the case that future evolution is in the process of collectively unfolding the yet higher stages of the Great Chain, as it has already unfolded the lower, it would give real religion—genuine spirituality and the deep sciences of the interior—an unprecedented role as the vanguard of evolution, the growing tip of the universal organism, growing toward its own highest potentials, namely, the ever-unfolding realization and actualization of Spirit.

Those, of course, are grand themes. Although there is considerable evidence to suggest that those themes are at least a genuine possibility (see *Up from Eden*), I would here like to focus on *individual* growth and evolution, because we can *already* do some profound and direct research on just this possibility. That is, we can already do extensive research on *the higher stages of individual growth and development*—the higher stages of moral growth, cognitive growth, affective growth, and interpersonal growth—using, of course, the deep science of the Upper-Left quadrant.

In fact, a good deal of orthodox research has already been done in each of those domains, evidenced in such ground-breaking research reports as *Higher Stages of Human Development* (Alexander and Langer), *Transcendence and Mature Thought in Adulthood* (Miller and Cook-Greuter), *The Future of the Body* (Michael Murphy), and *Beyond Formal Operations* (Commons, Richards, and Armon), and summarized in *A Brief History of Everything*.

The conclusion of all of this research is that, as we briefly saw in the last chapter, there are indeed several major stages of development beyond the formal-rational stage—higher stages of cognitive, affective, and moral development, among others. What we have here, in other words, is *the deep science of the higher stages of development or evolution in the Upper-Left quadrant*.

What remains to be done is to begin correlating this data with the simultaneous and corresponding *changes in the other quadrants*, thus generating an "all-level, all-quadrant" integral view. For example, what happens to brain physiology, neuro-transmitter levels, and the organic body itself when individuals move through these higher developmental stages? What types of worldviews are generated from these higher stages? How might these higher worldviews affect our political, social, and cultural institutions? If these higher stages are in fact stages of our own greater potentials, what types of integral techniques could facilitate this evolutionary growth? How will higher stages of growth affect our democratic institutions, our educational policies, and our economics? How will higher development alter the practice of medicine? law? government? politics?

In short, *how will these stages of our own higher evolution manifest in all four quadrants?* What higher art and science and morals await us? And what should we do about it now?

POLITICAL AWARENESS

Those are simply a few of the questions that we can ask, and perhaps begin to answer, with a truly integrative view. What I want to emphasize here is that this "all-level, all-quadrant" or *integral approach* is the direct result of the harmonization of premodern religion (all-level) with the differentiations of modernity (all-quadrant). And this integral approach forces us, as it were, to realize that any integration of science and religion is going to be much more than just that.

For example, what tends to be missing in most of the attempts to integrate science and religion is a deep discussion of its political dimensions. For modern science is part and parcel of the liberal Enlightenment and the differentiations of modernity, differentiations that brought with them the rise of the representative democracies, universal human rights, and the ideas of freedom and equality of all individuals, which in turn gave rise to everything from the abolition of slavery to feminism. Modern science was an integral part of this differentiated worldspace, in which those freedoms, values, and rights arose, and thus to talk genuinely and deeply of the integration of science and religion is to talk, sooner or later, of politics.

The core of the liberal Enlightenment was the assertion that *the state does not have the right to legislate or promote any particular version of the good life.* This can be put in several different ways: the state cannot legislate morality; there is a separation of church and state; the individual has the right to decide what constitutes his or her own happiness, as long as it does not violate the rights of others; the state may not unduly infringe on an individual's private life. These extraordinary freedoms—the product of differentiating the I and the WE—were part of the great dignity of modernity, of which modern science was an *inseparable* aspect.

This is why so much of the "new-paradigm" talk is profoundly off the mark. In the retro-Romantic versions, the dedifferentiation that is recommended would, if actually pressed into service, erase these freedoms and dignities. In the complexity theory versions, the political dimension is simply ignored (because monological science deals only with ITs, not with I's and WE's, and thus these "new paradigms" have nothing to contribute to politics). In the "quantum society" versions, the political and dialogical are rudely reduced to the monological, thus devastating precisely that which is claimed to be healed.

No, if there is to be a genuine integration of modern science and premodern religion, it will have political dimensions sewn into its very fabric. And just as the integration of modern science and premodern religion actually involved the integration of the differentiations of modernity with the Great Chain of Being, so the political integration of modernity and premodernity would involve the integration of the Enlightenment of the West with the Enlightenment of the East.

By the Enlightenment of the East I simply mean *any genuine spiritual experience*, whether of East or West. It is simply that the Eastern traditions have demonstrated, on balance, a somewhat more widespread reliance on a deep science of the interior, made most famous in Gautama Buddha's enlightenment under the Bodhi tree around the sixth century B.C.E. But any direct spiritual realization—East or West, North or South—conforming to the tenets of deep science could just as well serve as an example (Plotinus, Eckhart, Catherine of Siena, al-Hallaj, St. Teresa, Boehme, Rumi, St. Augustine, Origen, Hildegard, Baal Shem Tov, Dame Julian, etc.).

This spiritual Enlightenment is, by the virtually unanimous consensus of the higher sciences, the *summum bonum* of the Good life. And yet, by the tenets of the Enlightenment of the West, which must also be preserved, the state cannot in any way advocate or legislate in favor of this spiritual Enlightenment. The state must stay out of the business of publicly legislating the Good life, which belongs to the private sphere of each individual's own choice.

The only possible way to integrate these two demands is to realize that the *summum bonum* of the Good life lies not on this, but on the other, side of the political liberalism of the Enlightenment. That is, spiritual or transrational awareness is *trans*liberal awareness, not *pre*liberal awareness. It is *not reac-*

tionary and *regressive*, it is *evolutionary* and *progressive* ("progressive" being one of the common terms for "liberal").

Thus, genuine spiritual experience (or spiritual Enlightenment) as it displays itself in the political arena is not prerational mythic belief—which almost always wishes to coerce others to that belief, whether the belief be in God or Goddess, patriarchy or matriarchy, Gaia or otherwise—but rather transrational awareness, which, *building on the gains of liberal rationality and political liberalism*, extends those freedoms from the political to the spiritual sphere.

Thus, spiritual or transrational awareness accepts the general tenets of rational political liberalism (not prerational mythic reactionism), but then, within those freedoms, pursues spiritual Enlightenment in its own case; and, through the powers of advocacy and example, encourages others to use their liberal freedom—the Enlightenment of the West—in order to pursue spiritual freedom—the Enlightenment of the East.

The result, we might say, is a liberal Spirit, a liberal God, a liberal Goddess. In common with *traditional liberalism*, this stance agrees that the state shall not legislate the Good life. But with *traditional conservatism*, this stance places Spirit—and all its manifestations—at the very heart of the Good life, a Good life that therefore includes relationships in all domains, from family to community to nation to globe to Kosmos to the Heart of the Kosmos itself, by any other name, God.

(Traditional conservatism is in many important ways anchored in premodern worldviews—from mythic religion to civic humanism—whereas liberalism is largely anchored in the rational differentiations of modernity. Thus, the integration of premodern religion with the differentiations of modernity would open up the possibility of a significant reconciliation of conservative and liberal views. See *The Eye of Spirit* for further discussion of this theme.)

This is a "politics of meaning," to be sure, but a transliberal, not a preliberal, meaning. It does not come from the reactionary and regressive attempt to tell others what kind of mythology they must pursue. It does not claim that world transformation rests on accepting their paradigm. It does not attempt to heal the fragments by killing the contenders. It does not ask the state in any way to support or advocate on its behalf. It pursues none of those preliberal avenues.

Rather, standing *within* the political freedom—the liberal freedom—offered by the Enlightenment of the West, transrational awareness then moves into its own higher estate by pursuing spiritual Enlightenment, which it then offers, within that same political freedom, to any and all who desire to be released from the chains of space and time, self and suffering, hope and fear, death and wonder. In its own spiritual realization it is thoroughly transliberal, bringing together the Enlightenment of the East with the Enlightenment of the West.

Surely, both Enlightenments must be preserved. Both Enlightenments offer freedoms that took evolution billions of years to unfold. Both Enlightenments speak to the kindest heart and highest soul and deepest destiny of a common humanity. Both Enlightenments cry out to the best that we are and the noblest that we might yet become. Both Enlightenments taken together point to the liberation of all beings, both in the temporal realm (the Enlightenment of the West) and the timeless realm (the Enlightenment of the East), weaving together political freedom and spiritual freedom as the warp and woof of a culture that cares.

Could we really speak of world peace without both of these freedoms made available to all? Could any of us be deeply happy without these freedoms shining from the faces of all of Spirit's children? Could any of us truly sleep at night without all souls being liberated in this vast expanse? Could we dare begin to pray for ourselves without praying for one and all? And could any of us be truly free until all beings without exception swim equally in this ocean of emancipation?

And perhaps, political freedom joined with spiritual freedom, time joined with the timeless, space joined with infinity, we will come finally to rest, finally to peace, finally to a home that structures care into the Kosmos and compassion into the world, that touches each and every soul with grace and goodness and goodwill, and lights each being with a glory that never fades or falters. And we are called, you and I, by the voice of the Good, and the voice of the True, and the voice of the Beautiful, called exactly in those terms, to witness the liberation of all sentient beings without exception.

And on the distant, silent, lost horizon, gentle as fog, quiet as tears, the voice continues to call.

FURTHER READING

If you would like to pursue the topics raised in this volume, you might start with my books *A Brief History of Everything* and then *The Eye of Spirit: An Integral Vision for a World Gone Slightly Mad.* These books contain numerous references to other significant works in this area, and interested readers can begin following these leads as they wish.

SOURCES

Chapter 3: F. Crews, "In the Big House of Theory," *The New York Review of Books,* May 29, 1986. T. Kuhn, *The Structure of Scientific Revolutions,* 2d ed. I. Hacking, "Science Turned Upside Down," *The New York Review of Books,* Feb. 27, 1986. B. Barnes, *T. S. Kuhn and Social Science* (New York: Macmillan). D. Hoy, *The Critical Circle* (Berkeley: University of California Press, 1978). E. Gellner, "The Paradox in Paradigms," *Times Literary Supplement,* April 23, 1982.

Chapter 7: All quotes from C. Taylor, *Hegel* (Cambridge, Mass.: Harvard University Press, 1975). (See *Sex, Ecology, Spirituality* for an extensive discussion).

Chapter 9: R. Alter, "Review of *The Tunnel* by William H. Gass," *The New Republic,* March 27, 1995.

Chapter 14: All quotes from R. Lipsey, *An Art of Our Own* (Boston: Shambhala, 1988). For an extensive discussion of art and literary theory from an integral perspective, see *The Eye of Spirit,* Chapters 4 and 5.

Chapter 15: D. Matt, *The Essential Kabbalah* (San Francisco: HarperSanFrancisco, 1995). For a detailed discussion of an "all-level, all-quadrant" research agenda—and for my suggested answers to the questions raised in the section "Deep Science Research"—see *A Brief History of Everything* and *The Eye of Spirit.*

INDEX

absolute Self, 105, 112
agrarian societies, 47, 99, 100
Alter, Robert, 135
Althusser, Louis, 32
Anthropic Principle, 21, 22
Apel, Karl-Otto, 35
Aristotle, 70
art
  as the Beautiful, 49–50
  as cultural value sphere, 47,
    48–49
  differentiation from morals and
    science, 12, 47, 48–49
  dissociation from science and
    morals, 13, 55, 56
  predifferentiated, 48
  roots of modernity, 42
  use of "I" language, 50, 74
Asanga, 70, 80
Aurobindo, Sri, 18, 70, 103

Bauhaus, 43
the Beautiful, art sphere as, 49–50
  see also art
Bellah, Robert, 72, 96
Berdyaev, Nicholas, 6
Big Bang, 5, 20
Big Three, *see* cultural value
  spheres
biology, relationship to Great
  Chain of Being, 9
body, as level of Great Chain of
  Being, 7–8, 9
Bohr, Niels, 81

brain *vs.* mind, 62, 70–71, 82, 83
Brown, G. Spencer, 173
Buddhism, 5, 8, 21, 75, 112–13,
  171–72

Campbell, Colin, 96
category errors, 21, 141
childhood of humanity, 4, 16, 17
Chögyam Trungpa, 7
Chomsky, Noam, 148
Christianity, 5, 6, 16
Clausius, Rudolf, 81
cognition, integral-aperspectival,
  121, 131–33, 136
Coleridge, Samuel Taylor, 94
collective worldview, 71–72
  see also exterior-collective
    quadrant; interior-collective
    quadrant
communal outlook, 71–72, 156,
  157
Comte, Auguste, 16
confirmation/rejection
  defined, 156
  and direct spiritual experience,
    166, 172
  and scientific method, 144
  as strand of valid knowing, 156,
    159–60
consciousness, 105, 166
  see also eye of contemplation
construction, 104, 121, 123, 130
  see also interpretation
constructivism, *see* construction

ABOUT THE AUTHOR

KEN WILBER is the author of over a dozen books, including *The Spectrum of Consciousness, Grace and Grit, A Brief History of Everything,* and *The Eye of Spirit.* He lives in Boulder, Colorado.

ABOUT THE TYPE

This book was set in Berling. Designed in 1951 by Karl Erik Fors-
berg for the Typefoundry Berlingska Stilgjuteri AB in Lund, Swe-
den, it was released the same year in foundry type by H. Berthold
AG. A classic old-face design, its generous proportions and inclined
serifs make it highly legible.

Explore the world of Ken Wilber from Shambhala Publications at http://www.shambhala.com/wilber